Planet-Centered Astrology

Multiple Perspectives, Unified Vision

Stephanie Jean Clement, Ph.D, LPMAFA

Copyright 2012 by Stephanie Jean Clement
All rights reserved.

No part of this book may be reproduced or transcribed in any form or by any means, electronic or mechanical, including photocopying or recording or by any information storage and retrieval system without written permission from the author and publisher, except in the case of brief quotations embodied in critical reviews and articles. Requests and inquiries may be mailed to: American Federation of Astrologers, Inc., 6535 S. Rural Road, Tempe, AZ 85283.

ISBN-10: 0-86690-629-0
ISBN-13: 978-0-86690-629-6

Cover Design: Jack Cipolla

Published by:
American Federation of Astrologers, Inc.
6535 S. Rural Road
Tempe, AZ 85283

www.astrologers.com

Printed in the United States of America

Dedication

To my friends, business partners and teachers, Patricia Stauffer and Linda Black, to whom I owe more than I even know, and without whom I might never have embarked on my study of esoteric astrology.

And to my Higher Source.

Acknowledgements

There are too many people to mention here. Yet I must mention Mark Lipson, who first turned me on to planet-centered astrology, and Jeffrey Sayer Close, who made chart calculation of all the planet-centered views possible for both the MAC and the Windows environment.

Contents

Preface	xi
Introduction	xiii
Chapter One, Planet-Centered Astrology	1
Chapter Two, Mercury-Centered Astrology	19
Chapter Three, Venus-Centered Astrology	37
Chapter Four, Mars-Centered Astrology	51
Chapter Five, Jupiter-Centered Astrology	63
Chapter Six, Saturn-Centered Astrology	77
Chapter Seven, Uranus-Centered Astrology	91
Chapter Eight, Neptune-Centered Astrology	105
Chapter Nine, Pluto-Centered Astrology	117
Chapter Ten, Conclusion	131
Appendix One, Glyphs for Planet-Centered Astrology	135
Appendix Two, Moons of Planets and Two Asteroids Defined	137
Appendix Three, Houses in Planet-Centered Charts	139
Appendix Four, Eight Steps in Ritual Process	141
Appendix Five, Birth Data Sources	143
Glossary	145
Bibliography	147
Endnotes	151

List of Illustrations

Chapter One
    Geocentric Chart November 12, 2012
    Mars-Centered Chart
    Mercury-Centered Chart
    Venus-Centered Chart
    Jupiter-Centered Chart
    Saturn-Centered Chart
    Uranus-Centered Chart
    Neptune-Centered Chart
    Pluto-Centered Chart

Chapter Two
    Lincoln Mercury-Centered Chart
    Martin Luther King Biwheel—"I Have a Dream" Speech
    Barak Obama Mercury-Centered Chart
    Charles Darwin—Mercury-Centered Chart
    Charles Darwin Biwheel—Published *Origin of Species*
    Clarence Darrow Biwheel—Begin Scopes Trial
    William Jennings Bryan Biwheel—Key Scopes Trial Ruling
    Clarence Darrow Biwheel—Repeal of the Butler Act

Chapter Three
    Elizabeth Taylor Venus-Centered Chart
    Johnny Cash Biwheel—Recorded "Ring of Fire"
    Johnny Cash and Adam Lambert Venus-Centered Biwheel
    Johnny Cash and Adam Lambert Geocentric Biwheel
    Mata Hari and Greta Garbo Biwheel

Chapter Four
- Broncos and John Elway Geocentric Biwheel
- Broncos and John Elway Mars-Centered Biwheel
- Elway and First Super Bowl Loss Biwheel

Chapter Five
- Elisabeth Kubler-Ross Geocentric Birth Chart
- Elisabeth Kubler-Ross Jupiter-Centered Chart
- Elisabeth Kubler-Ross Book Publication Biwheel
- Sandra Day O'Connor Geocentric Birth chart
- Sandra Day O'Connor Samuel Alito Jupiter-Centered Biwheel

Chapter Six
- R. Buckminster Fuller Geocentric Birth Chart
- R. Buckminsnter Fuller Saturn-Centered Chart
- Tonya Harding Geocentric Birth Chart
- Tonya Harding Saturn-Centered Chart

Chapter Seven
- Arthur Young Uranus-Centered Birth Chart
- Brooklyn Dodgers Uranus-Centered Birth Chart

Chapter Eight
- Timothy Leary Neptune-Centered Birth Chart
- Timothy Leary Neptune-Centered Transit for Death Date
- Timothy Leary Neptune-Centered Transit for Date Ashes Sent into Space
- Jonas Salk Neptune-Centered Birth Chart
- Jonas Salk Biwheel for Vaccine Announcement Date

Chapter Nine
- Alice Bailey Geocentric Birth Chart
- Alice Bailey Pluto-Centered Birth Chart
- Alice Bailey Douglas Baker Biwheel
- Roy Campanella Geocentric Birth Chart
- Roy Campanella Pluto-Centered Birth Chart

Preface

Since I wrote *Planets and Planet-Centered Astrology* in the late 1980s and early 1990s, I have learned more than a thing or two about myself, about life, and about astrology. As I look back at that work, written largely on the basis of direct transmission and then researched in the crucible of my therapeutic practice, I realize I hardly knew anything then. Yet at that time I felt impelled to write about planet-centered perspectives.

Now we have the very refined capabilities of the Intrepid astrology program to provide extremely accurate planet-centered charts, and we have tremendous bodies of information about the solar system and beyond that were not in place in 1992.

Jeffrey Sayer Close has been developing his SELF-EVIDENT ASTROLOGY™ over many years, independent of my astrological work. His research concludes that astrology truly is self-evident if we examine the facts of astronomy.

Both Jeffrey and I were more than pleased, but not exactly surprised, when we found that our assessment of planet-centered charts and particularly the moons and asteroids associated with each planet, matched up on almost exactly a one-to-one basis. There is nothing quite like the treat of finding a person whose research has taken him down a different path, yet both of us arrived at precisely the same place.

The two of us are not alone in this. Mark Lipson created the Capella program that gave me my first glimpse of planet-centered perspectives. There are several other people in the U.S. and Europe who have studied planet-centered charts and presumably come to their own conclusions. I have given planet-centered charts to innumerable clients for their use, and they have more or less fallen in love with those mandalas that reflect essential facets of their beings.

Introduction

Astrology is founded on the premise, I believe, that the planets embody and exemplify archetypal concepts. The gods themselves, for whom the planets are named, were projections of human qualities onto deities. Thus all astrology reflects the human need to connect with universal themes and to understand transpersonal issues that impact our daily lives.

So why would you want to study astrology from the perspective of another planet? Until the latter part of the twentieth century, no one entertained any other astrological perspective than geocentric. Notice was given to the place a person was born, but the assumption was that since we live on the earth, the earth-centered viewpoint is the only one we can consider.

In addition, before the creation of the Digicomp computer for astrologers, other points of view could not be easily considered. The data used to develop the Digicomp was heliocentric to begin with, and the program translated that date to the geocentric view. Because the developers had the heliocentric data, Digicomp provided the means of looking at charts from the heliocentric perspective. While this availability did not set off an intense rash of exploration and introspection, it certainly captured the attention of notable astrologers like T. Patrick Davis and Phillip Sedgwick. Others have followed suit.

In the 1980s, Mark Lipson brought a Mars-centered chart to a class I was teaching. He had written a program to produce such a chart, and he continued to develop an MS-DOS program that utilized planet-centered viewpoints, including some or all of the moons of each planet. That class was my introduction to planet-centered astrology, and I have never looked back.

Now the Intrepid astrology program allows you to view astrological charts from the perspective of any of the planets, and these charts incorporate satellites of the planets. Jeffrey Sayer Close has developed the Intrepid program to run exactly the same on both Windows and MAC platforms, and he has utilized the most exacting mathematics to ensure accuracy of all the planetary and satellite positions.

Back to the question: Why consider the planet-centered perspective? The reasons to explore astrology from the planet-centered view are these:

1. Planet-centered charts provide direct expressions of each planet at your time of birth. The Mars-centered view, for example, shows the Logos (mind) of Mars when you were born. Planet-centered charts allow you to entertain a transpersonal perspective that is uniquely suited to your birth date and time. Only a handful of people on the planet will share identical planet-centered charts.

2. The earth-centered (geocentric) perspective is fraught with elements of ego (the Ascendant, Midheaven and personal planets). The planet-centered charts, being outside your ordinary awareness, contain far less in the way of ego involvement.

3. Each person identifies with his or her own planet-centered charts. I found this to be truly remarkable! In case after case, clients have looked at these charts and adopted them as their own. In fact clients, when cued to the energy of each chart, "get it" almost immediately.

4. Planet-centered charts offer clear choices/alternatives to resolve problems you experience in daily life. These problems can be delineated in the geocentric chart, but the solutions are not always clear. The planet-centered chart shows you a profoundly refined assessment of your best use of each planet's energy, and therefore shows you how to approach any problem you may have.

5. Planet-centered alternatives are not news to either astrologer or client. Your planet-centered charts indicate ways of acting in the world that you already understand and may simply have forgotten. This simple statement has been proven over and over again in work with clients.

6. Each planet becomes an ally in your quest for personal growth and spiritual understanding, and these allies speak to you in a language unique to you. Because of the speed of the moons, birth times have to be nearly exactly the same to produce identical charts. Even twins will note differences in the degree of the faster moons. For example, a ten minute interval moves Phobos forward about seven to eight degrees, and Deimos about two degrees. In the geocentric chart, a ten minute difference makes virtually no difference in planetary positions, and changes the angles only about 2½ degrees.

Jeffrey Sayer Close identified one particular celestial object associated with each planet. For Mercury and Venus, these upper harmonic objects are the asteroids Flores and Juno. For each of the other planets he found one satellite whose orbit and characteristics portray a "multiple" expression of the planet. These planetary companions are Deimos (Mars), Ganymede (Jupiter), Titan (Saturn, Miranda (Uranus), Triton (Neptune), and Charon (Pluto). These objects can be included in the earth-centered chart as a group and they will amplify your understanding of how each planet functions.

In the planet-centered portion of Intrepid, you choose the center planet and the program incorporates the appropriate moons. In the respective planet-centered charts the moons reveal the multiple expressions of that planet's energy for the moment of your birth. These additional satellites can also be added to geocentric charts. The moons of the planets reflect characteristics of the planet that you can use separately to make decisions and take clear actions that ideally suit your nature.

Modern society recreates the circle of helpers in fraternal organizations, secret societies of other kinds, and close groups of friends. We also explore multitudes of angels who surround us and offer protection and encouragement. You can think of the planets as your angels and the satellites as their emissaries—expressions of each planet's best wishes for your life.

It's important to remember that, regardless of the perspective, the planets are still the same planets as in the earth-centered chart, the signs are still the signs, and the aspects still work the same way. Earth's Moon is only included in earth-centered charts. The most powerful difference between geocentric and planet-centered perspectives, I believe, is that the planet-centered chart removes the emphasis of ego and provides a transpersonal view of the universe for each of us.

When you adopt a transpersonal perspective, you transcend the limitation of your personal, individual viewpoint or ego. The root "trans" comes from the Latin and means across, beyond, through, or change. One of the core principles is that the transpersonal perspective permits and indeed encourages you to change your thinking. "Personal," from the Latin *persona* means unique to the person or individual. Put the two meanings together and you get beyond the person, across personal boundaries, through the person, and change within the person.

Transpersonal perspectives have the effect of causing us to change what we know or believe about ourselves. I believe that change is essential if we are to overcome limitations of experience, both during the current lifetime or across the boundaries of the present lifetime.

Planet-centered perspectives allow us to get outside the geocentric (ego-centered) perspective so we can try on ideas unencumbered with personal history and emotions. Okay, we can't get away from personal history and emotions entirely, but we *can* consider other viewpoints. Planet-centered charts happen to provide each person with eight or nine alternative viewpoints. The content of each chart is the same except for the moons and asteroids associated with each planet. The moons and asteroids add one or more companions to elaborate or elucidate the nature of the center planet in your life.

As you read through the book and look at your own planet-centered charts, take the time to relax your mind and consider the alternative views your own astrology provides.

Chapter One

Planet-Centered Astrology

"The entire solar sphere is full of [planetary] bodies, characterized by the same features as are the seven and the ten, and each of them in some degree has an effect upon the whole, . . . There are more than 115 of such bodies to be reckoned with, and all are at varying stages of vibratory impulse. They have definite orbits, they turn upon their axis, they draw their 'life' and substance from the sun, but owing to their relative insignificance, they have not yet been considered factors of moment."—Alice Bailey, *Treatise in Cosmic Fire*[1], pp. 793-795

It is not always a simple matter to set aside egocentric activity in order to get an objective sense of the energy of another planet. The approach to planet-centered astrology in this book follows contemplative lines. You may want to use a similar approach in learning about these fascinating views of our solar system. First, try to get a sense of the planet and its energy, much as you might get information from the earth. Second, look at charts for family, friends, students, and clients to confirm your findings. Third, use meditation to gain a single-pointed focus of attention and increase mindfulness or awareness of what arises in meditation. Then review your insights to discover inconsistencies and to amplify ideas.

Each planet has a core interpretation in astrology. One basic premise of planet-centered astrology is that the planets are always the planets, and therefore interpretations of their meaning always retain the core of their essential message—they always reflect the same values or energy at a basic level regardless of the perspective from which they are viewed. Their spiritual essences do not change just because perspective changes. This fact provides a powerful reason why you will want to study planet-centered astrology. When you are able to approach each planet from eight or nine perspectives in addition to your earth-centered perspective, you then can refine your understanding of the planet's energy and its message in your life.

It's not only a matter of removing yourself from the ego-centered perspective of the earth—although that does help because you are no longer limited to the view from your own house. You can travel outward to experience different views of the solar system and the universe. In addition, you could say each planet has its own "ego." Each planet has a divine expression that it endeavors to radiate, consistent with the Mind of the Universe, as we experience it through the Mind of our own Sun.

The Sun is the source of all life as we know it, and is the center or focus of all activity. Each planet has a specific view of the Sun, the personal planets, and the other planets. Then each planet adds its own upper harmonics or companion planets to the picture.

Each planet-centered perspective adds a shade or nuance to the basic interpretation of the planet. To illustrate this fact, here is an example:

Let's look at Mars from various planet-centered perspectives using this randomly chosen birth data. We'll begin with the Earth-centered perspective.

About Mars in General

Your physical, emotional, mental and spiritual energy is shown by the placement of Mars in your chart. This energy covers the range from physical stamina to angry impulses to the passion of relationships and the compassion of spiritual involvement. The role of passion and compassion in the resolution of conflict cannot be over-estimated. Consideration of the potential direction of your energy in such situations is a valuable precursor to skilled action. The house placement of Mars indicates where your physical energy will be directed most strongly. It also shows how you use the power of devotion in your daily life. In this interpretation you discover that Mars signals the range of energy you experience in your daily life, and also the spiritual energy you have available.[2]

Once you have this basic meaning of Mars, you can examine its sign and house placement for a more focused perception of how you personally experience energy in your life. The house and sign placements add shades of meaning to the basic Mars interpretation. The Intrepid astrology software program provides keys to interpretation of every combination of astrological factors.

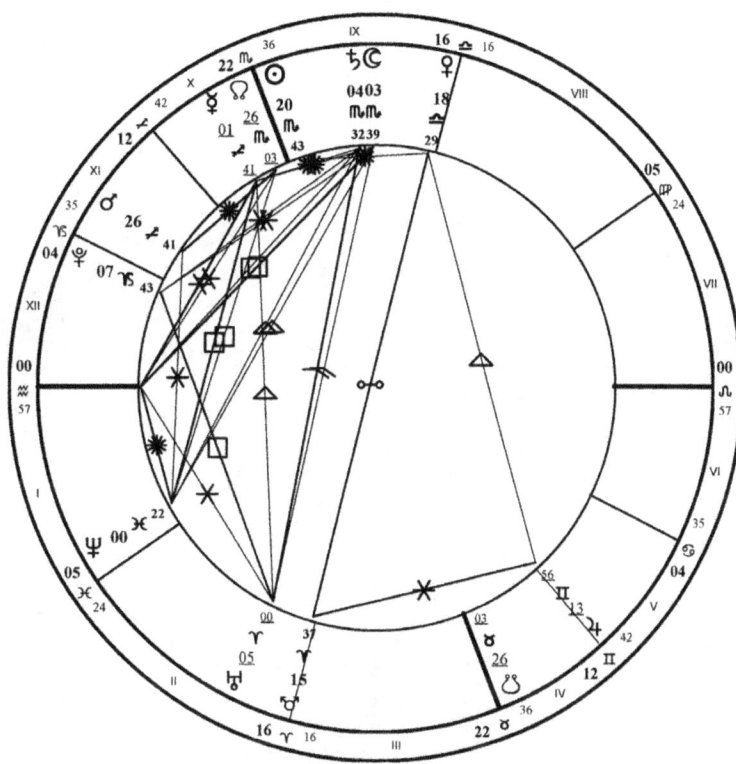

November 12, 2012, Washington DC, 12 pm EST
Mars' closest aspect is to the Nodes, suggesting that issues
surrounding energy may have karmic impact.

You will soon see that the keys follow an orderly pattern to extend your understanding of Mars. See the summary for a listing of the keys relevant to this chapter.

Earth-Centered perspective of Mars

Mars in Sagittarius
> *Your physical life focuses on the thighs. You use movement to keep your energy flowing. Your have a vigorous physical appearance; your mind is actively processing as well. You value the truth and expect it from yourself and others. This means that you frankly answer questions and expect the same from your friends and family. It is important for you to present your ideas clearly, in order to convince others. You love the challenge of games and other interactions. When planning vacations you tend to be very adventurous, straining the capabilities of the people you are with. When you have been overly extravagant, redirect your energy*

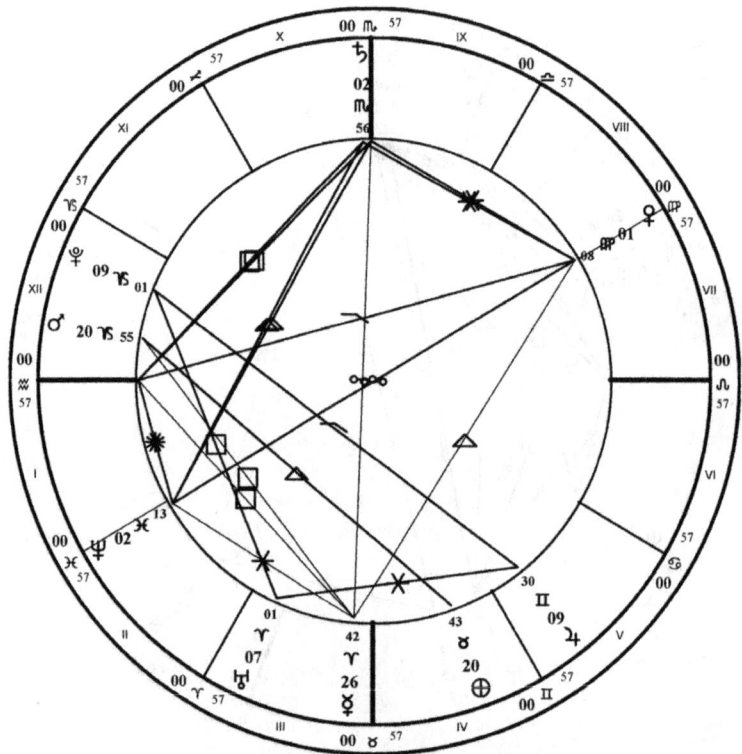

Heliocentric View; November 12, 2012, Washington DC, 12 pm EST

toward taking care of business, devoting yourself to improving the conditions around you without spending. When you feel you are becoming careless, devote your mind to a focused attention on the task at hand, leaving the next task for its proper time.

Mars in the Eleventh House

You are happiest when you are managing circumstances that come up in your daily activities. You enjoy social activities with groups but demand that you be allowed independence. Because your thoughts move so quickly, you are sometimes seen as inconsistent. You focus your energy on hopes and wishes and therefore get what you want more often than not.

With these three interpretations, you have the basic geocentric Mars meaning, Mars in a particular sign, and Mars in a particular house. You know pretty well how an individual with this combination will tend to act because you know where his or her energy is directed, both on the physical and the environmental level.

I have included a heliocentric chart for November 12, 2012 to complete the set of charts from perspectives other than geocentric. The present work does not explore the heliocentric chart. Three authors who have written on the subject are Philip Sedgwick (*The Sun at the Center*), T. Patrick Davis (*Revolutionizing Astrology with Heliocentric*), and Yamo Vedra (*Heliocentric Astrology or Essentials of Astronomy and Solar Mentality with Tables of Ephemeris to 1915*, first published in 1899, long before we had effective means to construct such charts).

I feel that the heliocentric chart speaks to one's mission. You could think of it as a view of your lifetime potential from the Sun's perspective. This chart can reveal areas of creativity you may have overlooked. It can show what your power and potential truly are, without limitations of ego and emotion. I have focused on the planets and their moons or companion asteroids, providing eight alternate views of your birth.

Now you can proceed to obtain eight more views of Mars.

Mars-Centered Perspective

Each planet-centered chart amplifies/multiplies the interpretation of the planet's energy. Thus the Mars-centered perspective reveals how energy is viewed from its own perspective, much as the earth-centered chart provides you a reflection of how you experience your immediate environment. Each planet-centered chart includes one or more companions. For Mars you have the upper harmonic Deimos and the second Martian satellite Phobos.

Phobos

Phobos, the faster moving satellite of Mars, indicates the active energy of Mars, while Deimos indicates the contemplative side. Integrating passion and compassion, Mars-centered charts define your unique capacity for demonstrating devotion, both actively and in thought. Phobos shows the active energy of creativity; it reveals passion in action. At an extreme, this satellite shows the compulsive behavior that agitates the personality.

Phobos in Capricorn

Acceleration of practical effort gets you ahead in career. Boldness makes your work extra creative and captures the desired attention. Your passion for action keeps you involved in physical activities. Creative action includes changing your mind and changing your direction.

Deimos

Deimos, the more thoughtful of Mars' two moons, indicates the obsessive thought associated with exaggerated emotions. The creative process of Deimos lies in

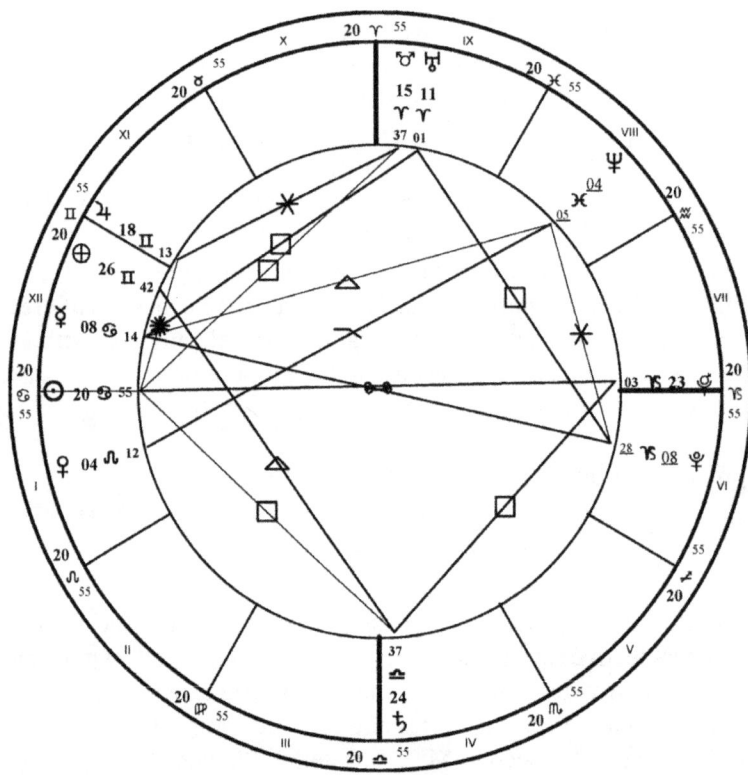

*Mars-Centered Perspective, November 12, 2012, 12:00 pm, Washington DC
The Sun in this chart, as in all planet-centered charts, is set on the Ascendant
to provide a uniform point of reference.*

thought; passion resides in the emotions rather than in action. Deimos can reveal, through its placement in the chart, your natural avenues for the expression of compassion. The fascination with a relationship is connected to Deimos. This moon indicates the nature of devoted mind for you as an individual.

Deimos in Aries

Desire drives your energy activity, including relationships. Inspiration comes from intuitive depths. Emotions generally drive you to achieve higher goals. Devotion to ideas eventually evolves into spiritual aspiration. Fascination with psychic insights keeps you on your toes mentally. Creative thought forms the basis for independent action. The passion of thought, for you, translates into psychical passion. Obsessive thought sometimes leaves you exhausted because you can't follow up on every idea.

The Earth from Mars' Perspective
The Earth's sign position in each planet-centered chart indicates how you engage in intelligent activity—that is, how you may use what you know skillfully to accomplish the goals of the planet at the center.

Earth in Gemini
You will relate to the energy of Mars through Gemini, and your awareness will focus on indecision and idealism. Intelligent activity can be a challenge at first, particularly when you have yet to make any big decisions. You approach every situation with huge optimism and you can chat with everyone from the tree trimmer to the president—and have fun doing it.

These interpretations provide a glimpse of the Mars-centered perspective. You immediately notice that the Earth in Gemini does not occupy the same sign as it would in the geocentric chart—opposite the Sun in Scorpio—in Taurus. This suggests that Mars would view the uses of energy somewhat differently than the person born on the date under consideration.

Interpretation of each planet-centered chart includes consideration of all the points in the chart and the aspects they make. The above example is just a taste of what you can learn from the Mars-centered chart, and you still have seven more planet-centered perspectives to consider.

Mars in Other Planet-Centered Charts

Each planet-centered chart provides a view of Mars from the perspective of the other planets. Lest you believe that these perspectives are useless to you, let's consider how you use this technique of changing viewpoints every day of your life.

When you hang a picture on the wall, probably you hold it up to get an idea of how it will look. Of course you are holding it, so you only get the viewpoint from about eighteen inches away. Maybe you have someone else hold it up so you can step across the room to see how it looks. You have changed your perspective.

When you are dealing with other people, you necessarily approach each situation first from your own self-centered perspective, and this includes physical, mental, emotional and spiritual perspectives. As you either try to figure out how to influence the interaction or how to cultivate compassion for the other person, you consider his or her viewpoints, and you try to do this on each of the four levels. With practice you get better at listening and perceiving what the situation feels like for the other person—you develop skill at considering other perspectives.

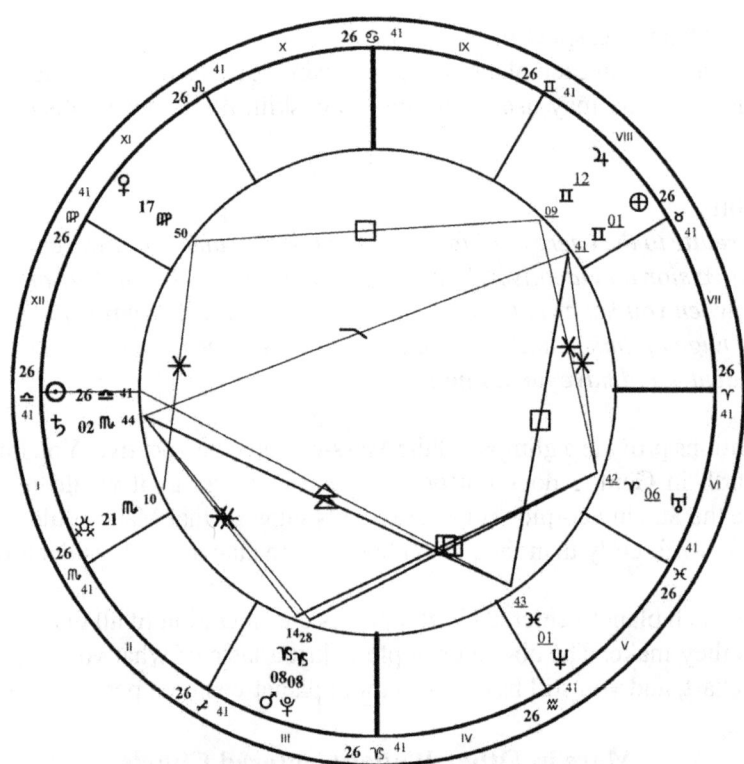

Mercury-Centered Perspective, November 12, 2012, 12:00 pm, Washington DC
Mars forms very close aspects to Uranus, Pluto and the Sun, suggesting that energy is an integral component of communication for the individual born at this time.

As you study astrology and study your own chart, you will benefit from the multiple perspectives of each planet through planet-centered astrology. I am reminded of a general problem-solving style that requires you to develop three distinctly different choices. You have to look at a problem from several viewpoints in order to define those choices. With astrology it is the same. When you consider different planet-centered perspectives, you gain a deeper understanding of the meaning of each planet in your life, and also the choices each planet offers.

Mercury Perspective: Mars in Capricorn

You have endurance that can be astonishing. You are often able to hang in there on activities when others have long since given up and gone home. This endurance is driven by your strong desire for success—success defined in your own terms.

Capable of serious planning and careful thought, you rarely make overt errors; you have thought everything out beforehand and worked through the kinks. Your

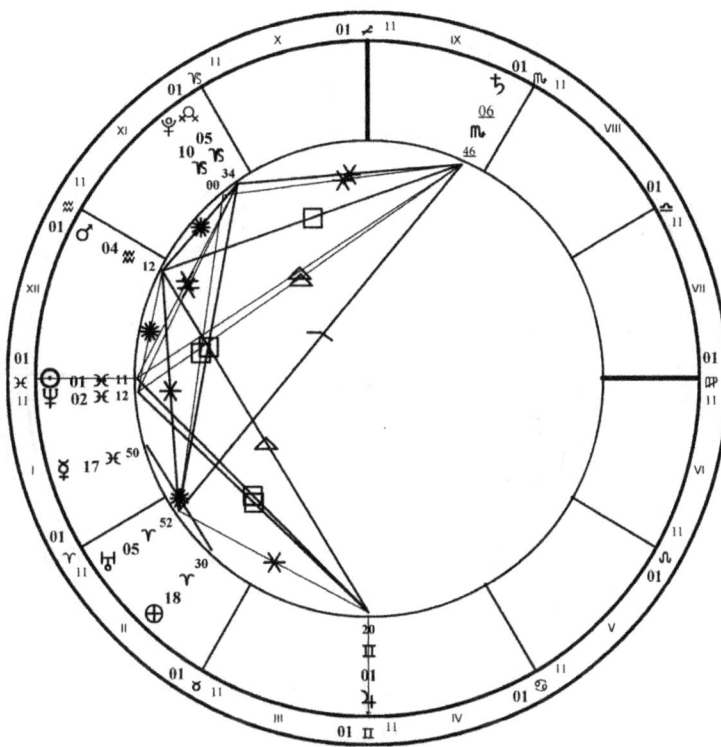

Venus-Centered Perspective, November 12, 2012, 12:00 pm, Washington DC
Here the closest aspect of Mars is to semi-sextile Juno, Venus' companion. Thus energy applied to humanitarian goals is focused and amplified.

ambition, coupled with an independent streak, makes you an ideal candidate to work for yourself. When you expect too much of yourself, re-focus your attention to the basics of what you need to accomplish. When you are caught up in your own ego, measure yourself against the scale of the mountain, material or metaphorical, which you intend to climb.

Venus Perspective: Mars in Aquarius

You arrange your thoughts and feelings on the basis of how they can support you in your work and play. Using unusual methods, you reorganize work to suit your sense of the human factors in the situation. This tends to be rather impersonal as your focus is more global in scope. Not one to shy away from controversy, you use your energy to logically think through problems. When you jump into reform without thinking it through thoroughly, recall your devotion to the overall human condition in order to regain the larger perspective. When you are inconsistent in your

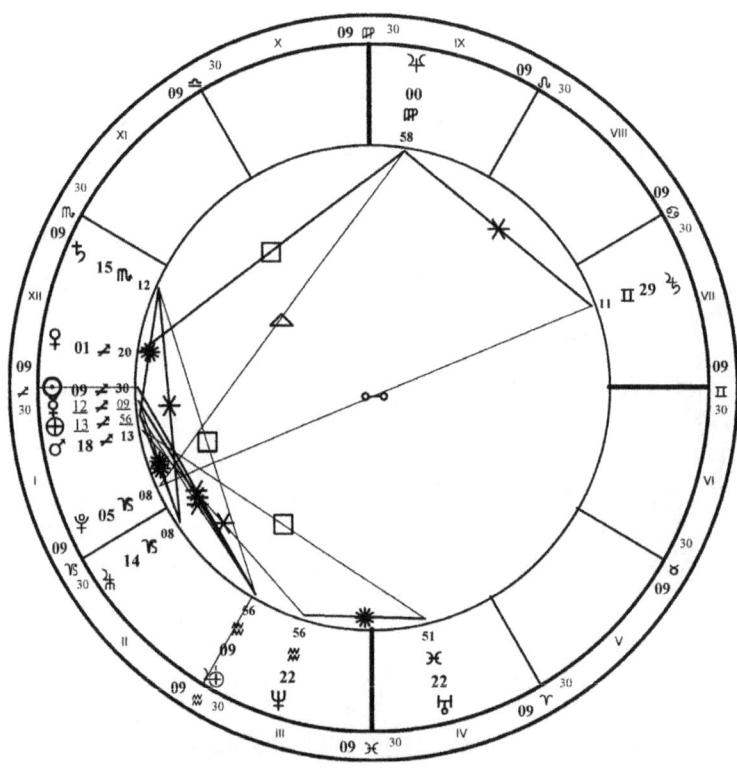

Jupiter-Centered Perspective, November 12, 2012, 12:00 pm, Washington DC
Note that the inner planets, including Mars, begin to bunch more tightly together because the viewpoint is so distant from the Sun.

activities, exercise the mental deliberation that serves you well in your work. Seek logical support and constructive advice.

Jupiter Perspective: Mars in Sagittarius

Success comes when your mind is actively processing. You value the truth and expect it from yourself and others. This means that you frankly answer questions and expect the same from your friends and family. It is important for you to present your ideas clearly in order to convince others. You love the challenge of games and other interactions. You can direct your energy toward taking care of business, devoting yourself to improving the conditions around you. When you feel you are becoming careless, focus your attention on the task at hand, leaving the next task for its proper time.

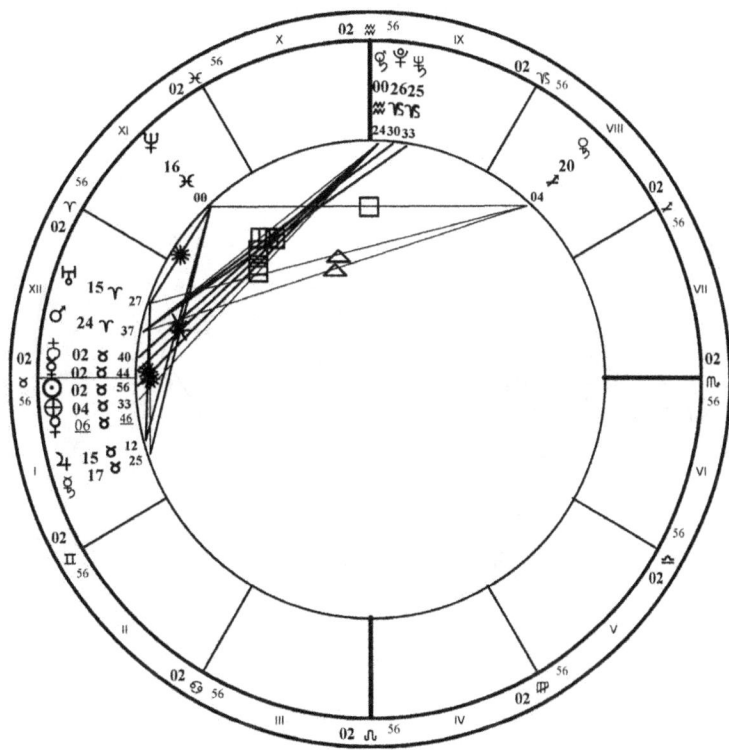

Saturn-Centered Perspective, November 12, 2012, 12:00 pm, Washington DC
Within one degree of the square to Dione, Mars energy challenges and tests structures,
at least from Saturn's perspective.

Saturn Perspective: Mars in Aries

The urge to action drives you. You live for independence and will actively seek to maintain your own and that of others as well. There is a tendency to waste energy through indiscriminate action or impatience. Your sexual activity depends directly on the ideas you have about it. For you, planning is necessary to get the most out of intimate relationships. Otherwise the fun is over before you are ready to start. When you feel a lack of energy, apply yourself to rest the way you apply yourself to activity. Become devoted to a sense of wellness.

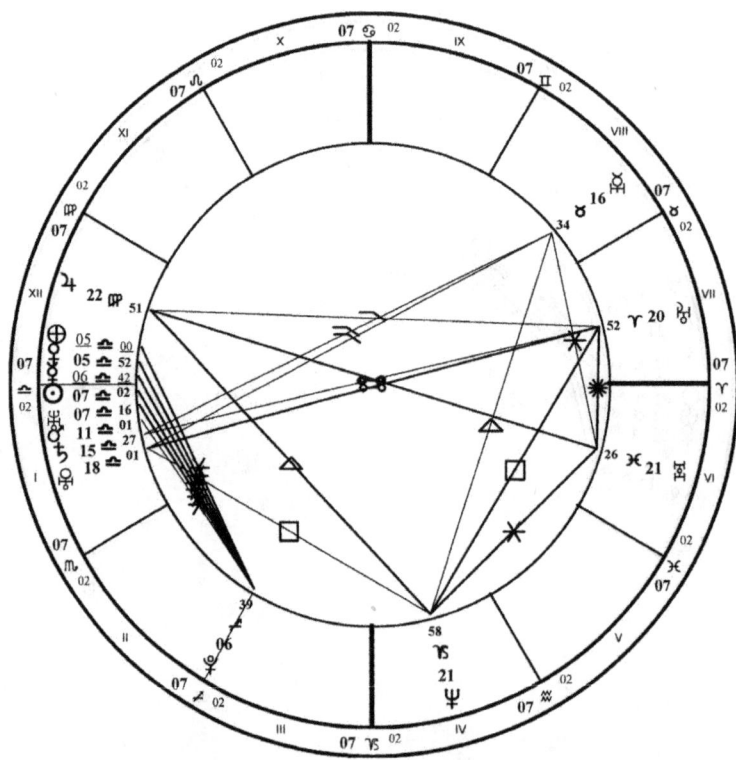

Uranus-Centered Perspective, November 12, 2012, 12:00 pm, Washington DC
Mars, although part of the bundle of planets around the Sun, forms no tight aspects.

Uranus Perspective: Mars in Libra

Your energy tends to come in bursts, usually connected to action in the environment, good or bad. You respond to outside impulses by throwing yourself into activities. This strategy works well most of the time, but occasionally you find yourself near exhaustion from all those interactions with other people. When this occurs, take time for contemplative activities to restore your energy. You enthusiastically engage in teamwork projects, relishing the social contact. When you are driven by moods, use your devotion to the project at hand to get yourself back on track.

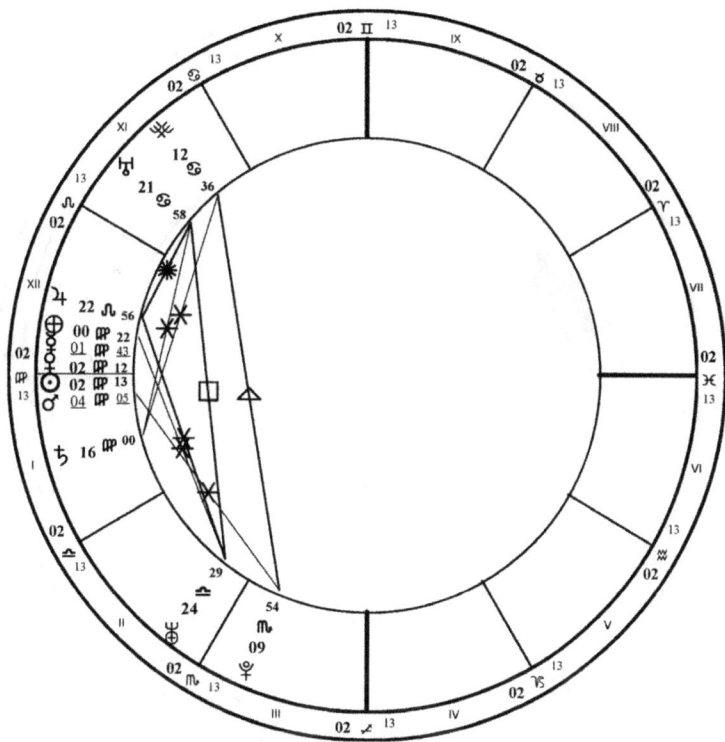

Neptune-Centered Perspective, November 12, 2012, 12:00 pm, Washington DC
In this chart Mars retrogrades (note the underlined numerals that indicate the retrograde) toward the Sun, indicating that solar and Martian energies are strongly blended.

Neptune Perspective: Mars in Virgo

Gutsy in all your activities, you operate from a position of fluid effort. Your physical power is centered in the abdomen and your psychic awareness comes from this balance point as well.

You are able to work with every situation, intuitively uncovering the meat of the activity. Where others have missed the point, you are working methodically through the details of projects or activities with your finger on the pulse of what is most important. If you are overly critical of others, recall your devotion to the outcome and employ a constructive process.

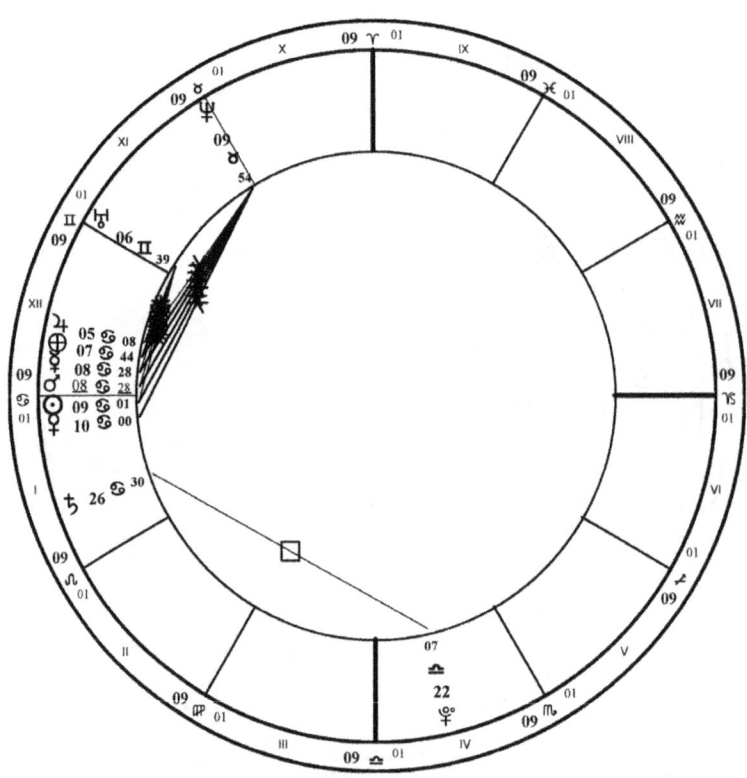

Pluto-Centered Perspective, November 12, 2012, 12:00 pm, Washington DC
With Mars exactly conjunct Venus (we expect them to be close), an individual born at this time can expect powerful psychic forces in romantic relationships.

Pluto Perspective: Mars in Cancer

Your energy focuses on home and family. For you, nurturing relates not only to food but also to the psyches of family members. You have an intense emotional life that can be cultivated through full-flavored relationships with each of them. You tend to operate at an instinctive level for the most part, going on feelings instead of facts. The psychic vibrations in your home provide clues about how to nurture household members. Effort in relationships rewards you well. Children accept your psychic insights without question.

Aspects in Planet-Centered Charts

While the aspects retain their character in planet-centered charts, the additional points (earth, asteroid or moons) present new factors for consideration. Aspects to the planetary companions provide a window into the psyche. Because the planetary moons can be anywhere in the chart, all aspects can form between them and other planets, providing a complete range of potential. Aspects to the planetary moons often feel emotionally charged because they probe unconscious territory and reveal archetypal themes of particular importance to the individual.

The following list of aspects addresses themes that may emerge in planet-centered charts with intense clarity:

- Conjunction—The conscious and unconscious minds work together. The separation of the two may be small in terms of awareness yet profound in terms of quality.
- Semi-sextile—Growth in understanding of the relationship of the conscious to the unconscious can be painfully poignant. While the adjacent elements in the semi-sextile aspect are traditionally considered to be incompatible, they in fact represent natural alchemical connections.
- Semi-square—Tension between the conscious and unconscious minds heightens the sense of being. Choices between conscious and unconscious perspectives becomes very important.
- Sextile—The opportunity is to achieve a dialogue between formerly warring parts of one's being. The conscious can feed the unconscious and vice versa.
- Quintile—A creative relationship can develop, with movement between the conscious and unconscious as the reward.
- Square—The challenge is to balance the conscious and unconscious minds. The sign of the Earth defines what is most easily made conscious; the sign of Venus defines what needs to become more available.
- Trine—There is an easy relationship between the conscious and unconscious elements of the mind. This aspect indicates a person who is comfortable with the mind—the ego has mediated between the inner and outer awareness.
- Sesqui-square—Agitation occurs, particularly when formerly unconscious material pushes its way into the awareness. One is alarmed by the force of this action.
- Bi-Quintile—The interaction of conscious and unconscious leads to highly organized creative process in which alchemical movement occurs freely.
- Quincunx—Adjustment of one part of the mind to another takes place in order to promote growth and understanding. To the extent that such growth is expected and accepted, the adjustment will be relatively painless.
- Opposition—Awareness of the values of the conscious and unconscious minds is in the

foreground of this individual's experience. There may not be understanding of how the two can ever come together; yet there is an appreciation of the value of each.

Summary

The table included here shows the center planet, relevant positions, and keys. You will immediately see that every key addresses the central focus of Mars—energy—and incorporates a clue to the specific perspective in each case.

Views of Mars, November 1, 2012, 12:00 pm, Washington DC

Center	Mars Position	What Is Energized
Earth	Mars in Sagittarius	Intellect
Earth	Mars in the Eleventh House	Citizenship
Mars	Phobos in Capricorn	Leadership transformations
Mars	Deimos in Aries	Births
Mercury	Mars in Capricorn	Ideas about leadership or patriotism
Venus	Mars in Aquarius	Coming together of humanitarian activities
Jupiter	Mars in Sagittarius	Growth of intellect
Saturn	Mars in Aries	Integration of birth
Uranus	Mars in Libra	Sudden change of religion or philosophy
Neptune	Mars in Virgo	Reversal of industriousness
Pluto	Mars in Cancer	Unexpected events with home, children

The traditional Earth-centered interpretation focuses on the individual's intellectual energy and how it may be used to further interests in citizenship. Each additional key reveals how this compelling urge may be enhanced, using the energies of the other planets. As you read your own chart, you can then examine the aspects Mars makes to identify which of the items in your personal list of choices may be easier to implement, and which may take a bit more practice or thought. None of the listed keys for your own charts represents an alien perspective. Rather, you will find that you identify with each of them, either immediately or after you have studied the keys and interpretations.

The following chapters take each planet in turn. The discussions include information about the nature of each planet, associated mythology and psychology, key words for the planet's expression, astronomy of the planet-centered perspective, and the role of the asteroid or moons associated with the planet. Examples have been chosen to illustrate the dynamics of each planet's perspective.

Chapter Two

Mercury and Mercury-Centered Charts

As far back as Sumerian mythology, and perhaps even farther, Mercury was seen as the messenger of the gods, indicating this planet's position as the representation of archetypal communication. Throughout the Bible, mythology, Tarot, and alchemical writings, we find communication as a central theme for human life. Both testaments of the Bible begin with statements that call upon the energy of language for the act of creation. The Old Testament begins with these words: "In the beginning, God created heaven and earth. . . . And God said, Let there be light. And there was light." The word of God, His communication, was the first step in creation. The first words of the Gospel according to John read: "In the beginning was the Word, and the Word was with God, and the Word was God." John shows us that the Word existed before anything else in the world, and that this Word is synonymous with God.

Human society depends on effective communication. We have memory, forethought and speech. Astrologically, all tools and means of communication, including travel, have the essence of Mercurial archetypal energy.

Mythology of Mercury

True to his double nature, Mercury represents the fusing of beliefs from various cultures. The Egyptian Thoth carries a magical tool very similar to the caduceus of Mercury. The symbol of medicine, this tool merges Mercury's roles as messenger of the gods and lord of science. The role of magic is retained because medicines often seem to have magical healing powers. The Romans infused Mercury into every culture they conquered, providing local deities with dual names by including Mercury as the first name and retaining the local name (a habit the Romans had with all their gods).

Norse mythology speaks of Odin or Woden, a god who presumably breathes life into people. Mercury, as ruler of respiration and inspiration, governs life as well as mental inspiration. Woden and Mercury share the name of the day of the week in European languages: Woden for Wednesday, Mercury for Mercredi or Miercoles (French and Spansih).

Woden is also seen as the god of war, but if this be the case, it is because he has control over magic, another of Mercury's general attributes. The first card of the Tarot, the Magus or Magician, represents will and dexterity. The magician has all the tools of the material and magical world: cups, swords, staffs and coins are representative of the four basic elements; the magician has all of these at his disposal for his work. Magic is a mercurial theme in many cultures.

The number one represents synthesis, as "nothing can be thought of without parts. It is the universal principle of existence, the creative intelligence of Deity, the motive force of the universe which in man becomes will. In the macrocosm [the number one] stands for unlimited potentiality, and in man for relative potency."[3] Synthesis of deities from different cultures underscores the oneness or Unity inherent in the energy of the planet Mercury in astrology.

What is the relationship between Mercury and the elements that comprise the world? Mythical Mercury is capable of using the elements to shape our world; we as human beings have this capability by right of our mental faculties. We use the seemingly magical ability of creation and vitalization of mental images to produce change in our outer and inner worlds, and change begins with communication.

Dane Rudhyar, an exponent of transpersonal astrology, states that "Mercury refers on the one hand to the most basic operations of the solar vital energy—the organismic power of integration—and to mental processes."[4]

The blending of high-minded and idealized concepts focuses on the alchemical quality of Mercury. Mercury moves so quickly that its transits seem ephemeral, yet the saying, "Haste makes waste," suggests that all is not entirely wonderful where Mercury is concerned. Our minds operate at such speed that we are forced into "forgetting" much of what we experience in order to make any sense of what we consciously recall. This sets up the possibility of "forgetting" anything that is unnecessary to our immediate needs, anything that is painful and uncomfortable for us to consider, in fact anything that we choose not to remember and deal with in consciousness.

We repress or suppress information that we cannot handle, forcing it into the unconscious. Later, we may find the unconscious overcrowded with uncomfortable material and we have to bring it into consciousness for review. Once again, Mercury is the agent of this process, only now he is the mediator between conscious and unconscious, an internal process of communication, where before he was the avenue of communication with the outer world.

"Alchemically, Mercury represents the philosopher's stone. This touchstone of alchemy is Truth; for when truth is pressed against anything its eternal principles are revealed, and these all enduring qualities thus obtained constitute the gold of their underlying nature. . . . Thus truth is a freeing and transmuting power, a feeling as well as an intellectual perception."[5]

If we experience confusion, anger, depression, or any other emotionally colored picture, Mercury serves as a therapist, helping to identify the truth of the matter. He can also be an inner source of wisdom. The alchemist employs four basic processes of *calcinatio* (fire), *coagulatio* (earth), *sublimatio* (air), and *solutio* (water) in his work. Whatever favorite definitions of the elements we may choose, it is the function of Mercury to make use of them to establish communication in our inner and outer lives.

Mercury relates to the signs Gemini and Virgo, identifying the factors of communication and intelligence in the ordinary workings of our existence. Gemini is the communicator; Virgo is the analyzer. On the spiritual path, according to Alice Bailey, Mercury relates to Aries. The creative impulse of Will is emphasized here. "Harmonizing the cosmos and the individual through conflict, producing unity and beauty . . . the birth pangs of the second birth," these are the ways in which Mercury can work.[6] We find ourselves changing and changed by our interaction with Mercurial energy.

On the hierarchical level of esoteric astrology we find Mercury related to Scorpio. Traditional astrology clearly defines Scorpio as a sign representing profound changes—birth and death among them—and the ruler is Pluto. In contrast, on the spiritual path, Mercury, representative of fourth ray energy—harmony through conflict—reflects the precise nature of the alchemical process; Mercury is the inspiration for the process, the guide through the process, and the result of the process all in one.

The multiple aspect of Mercury can be summarized as follows:

1. He consists of all conceivable opposites within Unity.

2. He is both material and spiritual.

3. He is the process by which the lower and material is transformed into the higher and spiritual, and vice versa.

4. He is the devil, psychological guide, evasive trickster, and God's reflection in physical nature.

5. He is the shaper of worlds.

As such, he represents on the one hand the self and on the other the individuation process and, because of the limitless number of his names, also the collective unconscious.

Mercury, traditionally connected with communication in astrology, is the primary and most effective tool we can have for working with ourselves and others on any level of consciousness. In daily chart reading and counseling we find that Mercury is a solid indicator in a birth chart of the method the individual uses to process information and communicate with others. Aspects between Mercury and other planets show us the relative ease or difficulty of communication, the houses show the areas of life that are most easily communicated, and the signs show the quality of communication. Transits and progressions to Mercury indicate times when communication will be influenced for good or ill.

Mercury can also reflect the energy of other planets that in turn form their own difficult aspects. Mercury is the avenue for information to pass from one individual to another, and from the unconscious into consciousness; Mercury is the vehicle, not the message

Mercury Key Words

Physical	*Emotional*	*Mental*	*Spiritual*
senses	restless	reason	Buddha
awareness	nervous	analytical	healing
motor nerves	anxiety	scientific	consciousness
speech, hearing	conflict	scientific	revelation
	relationship	expression	Planetary Mind
			intuition

Mercury-Centered Charts

To begin considering the Mercury-centered perspective, you must understand some unique astronomical facts about Mercury:

1. The orbital eccentricity of Mercury results in the most elongated elliptical orbit of any of the planets except for Pluto. While zero eccentricity would be a perfect circle, Mercury's eccentric-

ity is 0.205. This eccentricity results in a corresponding variation in speed, from 38.7 km per second at aphelion to 56.6 km per second at perihelion (closest approach to the Sun).

3. Perturbations in Mercury's orbit move the point of perihelion very gradually, about 43 seconds of arc per century. This movement confirms Einstein's theory about the curvature of space around large masses. Perihelion occurs in the sign of Taurus (midnight at 0 degrees of Mercurial longitude) at one orbit and in the sign of Scorpio (noon on the following orbit of the Sun).

4. Mercury's polar axis is perpendicular to its orbit, resulting in no seasonal variations.

5. Mercury's rotational period has a 2:3 relationship to its orbital (sidereal) period.

Perhaps the most unusual effect of Mercury's rotational/orbital relationship is that observers on the planet would see erratic solar movement that varies according to the longitude of the observer. For example, at 90 or 270 degrees, the observer would see a double sunrise or double sunset at perihelion passage. The following perihelion would result in the opposite (double sunset or double sunrise). An observer at 0 or 180 degrees would see a solar motion not unlike what the Earth observer sees when planets retrograde—the Sun would rise for about 1.5 Earth months toward the Mideaven. Just before noon the Sun would loop back on itself, much as retrograde planets appear to do, then stop and move forward in the sky again.

This motion depends on the relationship between the Sun and a particular longitude on Mercury's surface, whereas retrograde motion from the geocentric viewpoint depends on the relationships between the Earth and another planet in their respective orbits.

Consider the Sun's apparent reluctance to move forward from a personal point of view: if the largest psychological object in childhood (the parent) hesitates, the child might feel ungrounded and insecure. You might doubt the correctness of instructions from your parent if he or she were to vacillate and contradict decisions. If your parent stops to make sure you understood, then you feel more secure.

Mercury-centered charts can be understood in terms of the powerful child/parent relationship with the Sun, as the Sun dominates the Mercury heavens like no object in Earth's sky. On Mercury the Sun appears to be about three times as large as the geocentric Sun. Tremendous heat is generated at mid-day due to the slow rotational period of Mercury (the Sun will have been in the sky for nearly 30 Earth days). This is because Mercury rotates on its axis once in 58 days. The speed of the god Mercury in his errands among the other gods was a mystery even to them, so it should not surprise us that we have to "bend" our minds to grasp the intricacies of the Mercury-centered viewpoint.

More Mercury Mythology

The Egyptian Anubis conducted souls to the underworld. Thoth, another Egyptian deity, was the god of wisdom, Lord of Words, and inventor of writing. The Egyptians taught that Thoth was responsible for the whole of creation, accomplishing the task through words alone.

Hermes was the Greek god who gave Pandora the ability to lie. He was considered to be a god of travelers and of commerce and he had magical abilities too. On the day he was born, Hermes began his life of pranks by stealing Apollo's cattle. He continued in this vein but gained the sympathy of all the gods anyway because he was helpful and ingenious.

The Trickster appears in myths worldwide. He (or she) throws information at us from unexpected places to get us to shift our attention. Yet the Trickster always has something constructive in mind, if we can just decipher it.

Alchemical Action

The extreme temperature changes on Mercury result in intense alchemical processes: extreme *calcinatio* (burning) is followed by utter *coagulatio* (solidifying), repeated during each "day" cycle, or 176 Earth days. Thus an observer on Mercury would alternate between *calcinatio* and *coagulatio* approximately twice each Earth year.

In examining Mercury-centered charts in comparison with geocentric, we find that Mercury is stationary in the geocentric chart at the same time Earth is stationary in the Mercury chart. This suggests that delineation of the meaning of stationary positions has a certain validity from both points of view. For example, a Mercury station in the geocentric chart may involve the receipt of documents or other communications. At the same time Earth is stationary in the Mercury-centered chart, making it easy for Mercury to locate Earth and aim a message toward it.

Communication from Mercury is not always just what we want to hear. It is, however, just what we need. Many Mercury aspects signal the happiest events of our lives; a few Mercury aspects bring with them the hardest and best lessons we will ever face.

Mercury and the Zodiac

Paul Foster Case said this about the elements:
> Four are the subtle principles
> Which the wise conceal from the uninitiates by the names:
> FIRE, WATER, AIR AND EARTH.[7]

Why would the wise conceal the fundamental nature of the universe from so-called uninitiates? Is there something so mystically or psychically powerful that the general public should not know about it? In the twenty-first century so many resources exist to expound the meaning of these four basic elements that we can no longer think of them as secret. Ancient wisdom does suggest that some principles are self-secret—that is, even when people read or hear about them, understanding is fleeting or non-existent. The meanings of the elements, the planets, and all of astrology are available to anyone who seeks to understand them. There is a mystery, but no secrets.

The esoteric principle of harmony through conflict is carried through the action of the four elements or principles; all major traditions utilize this basic array of energies, sometimes describing a fifth factor, ether, all-encompassing space or the void. Mercury, because of its nature as mediator and communicator, makes direct use of the elements or energies in its action.

Mind in the Mercury-Centered Chart

Starting with the premise of esoteric astrology that the solar Logos (Sun) is the point of manifestation for life, and that the solar system is the body or vehicle for that manifestation, energy radiating from the Sun meets the planets and activates their energies in specific ways. Each planet has a logos or mind of its own. Our actions reflect our understanding of the planetary minds and the mind of the Sun as well. Astrology provides elegant tools for understanding the solar mind, planetary minds, and human minds.

Mercury-centered charts provide a transpersonal (outside the personal) perspective, showing what we can do with Mercurial energy and how we can most effectively proceed. These charts also provide a specific focus on the intention of the Sun. In esoteric astrology Mercury has these attributes:

- Mercury indicates how we may achieve wisdom, enlightenment, or self-realization.
- Mercury represents potential healing for ourselves and others.
- Mercury expresses through the abstract mind. Therefore Mercury's greatest power becomes available after the first Saturn opposition in the geocentric chart—the approximate age when abstract thinking develops.
- Mercury in Gemini reflects the kinds of conflicts that occur when we attempt to learn new information or undertake new activities. The struggle is primarily between mind and the material environment.
- Mercury in Virgo conditions our minds so that we may evolve. There is a struggle in this effort as well, but here the struggle is between mind and spirit.
- Mercury in Scorpio reflects the power to transform perceptions, purify your own vision, and transcend the world's apparent limitations altogether.

The Mercury-centered chart addresses esoteric principles in familiar ways. The signs, planets and aspects are essentially the same as they are in the geocentric chart. However, the focus is different. From Mercury's perspective, the mind is the pivotal point in manifesting or expressing your true mission. Each planet's placement in the Mercury-centered chart addresses core issues of language and communication because language is a powerful vehicle for developing the mind.

In addition, the Mercury-centered chart reflects your unique approach to Unity. As you examine the planets—the players in the Mercury-centered chart—you learn how different energies come together to aid your quest for understanding of the world and your place in it. The planets also provide perspective concerning the expression of solar energy. The Sun in any planet-centered chart reflects your life's mission. In the Mercury-centered chart, the Sun indicates how your mission benefits when you use all the faculties of your mind.

Mercury's Planetary Companion

Mercury and Venus have no moons. Their planetary companions have been found among the asteroids. Mercury's companion is Flores. If Mercury is lord of mind and logic, then Flores is both memory and the capacity for expression. If Mercury is the lord of words, then Flores is the reporter/recorder who tracks his remarks and actions. Flores also acts as an assistant to the magical Mercury, creating just the right balance of distraction and attraction to cause harmony to reign at the end of every effort. If Mercury is the Trickster, the Flores is his capable assistant, distracting us in meaningful ways while Mercury works his magic. Then Flores announces the results in no uncertain terms.

Words, magic, and results add up to the way intelligence acts from Mercury's perspective. Flores is like a trained observer who captures both cosmic and individual essence of communication. The placement of the asteroid Flores and its aspects weave a complex story about your capacity to communicate and to express the knowledge you have gained.

Mercury-Centered Examples

Three individuals represent the power of the written and spoken word to change the course of history: Abraham Lincoln, Martin Luther King, Jr. And Barak Obama. Interestingly, Charles Darwin, a time twin of Lincoln, set the course for another chain of events that rocked the world of science. Both Lincoln and Darwin had dramatic impact on the world through their words, so the Mercury-centered chart is definitely a good place to examine some of their accomplishments and legacies. The Mercury-centered charts reveal connections that you may not have considered!

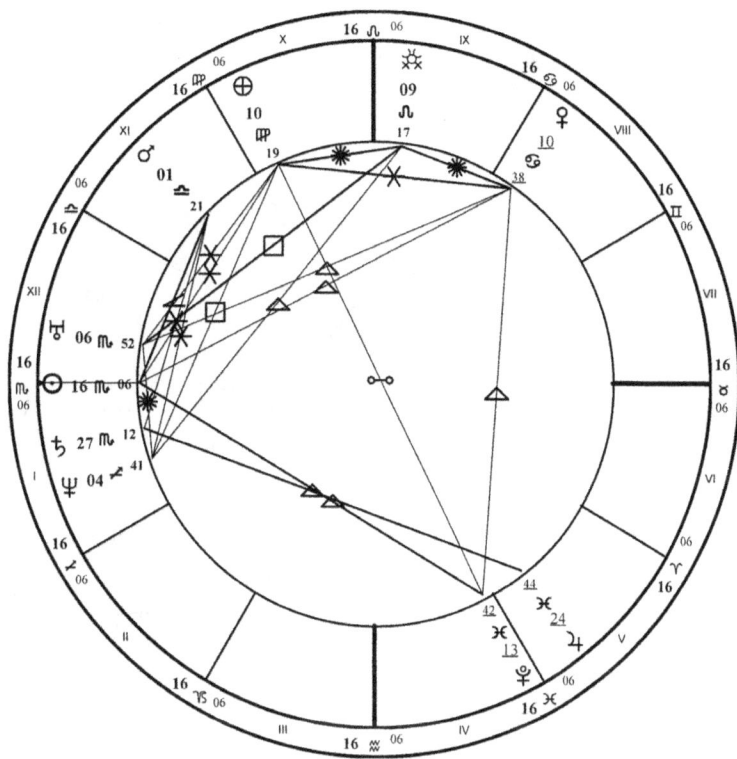

Mercury-Centered chart for Abraham Lincoln, Time Twin of Charles Darwin
Birth Data: February 12, 1809, 6:54:06 am LMT, Hodgenville, Kentucky

Abraham Lincoln
Every American child has heard the Gettysburg Address, and most had to memorize it as well. This speech defined Lincoln's sentiment concerning the Civil War in two minutes and fewer than 275 words. Most of you probably don't recall the name of the featured speaker, Edward Everett, who spoke for two hours. Lincoln began the speech with a sentence that refers to the equality of all people, and ended with a dedication to those who fought for that equality.

Lincoln's Mercury-centered chart features a Kite pattern that includes the Sun, Pluto, Venus, and Earth. Flores, Mercury's companion, sits in the ninth house in Leo, immediately suggesting that Lincoln will be famous for his thoughts and words. The pattern reflects the effort Lincoln put into his work, from the time he studied completely on his own to his role as president.

The Mercury-centered chart carries forward after Lincoln's death, to impact the lives of all Americans.

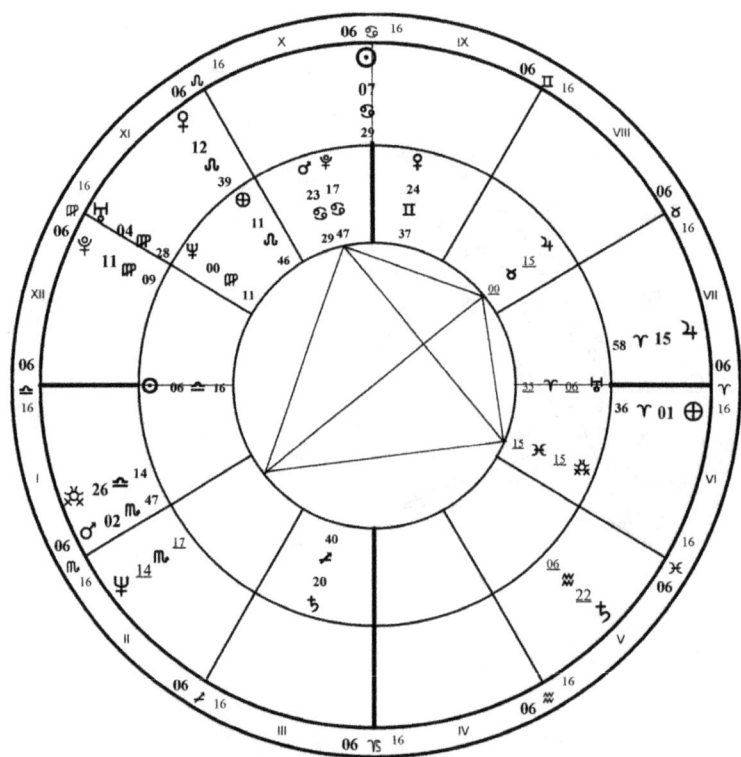

Mercury-Centered charts for Martin Luther King (inner) and his most famous speech (outer)
Birth Data: January 15, 1929, 12:00 pm CST, Atlanta, Georgia
Speech Data: August 28, 1963, approximately 5:00 pm, Atlanta, Georgia

I Have a Dream Speech

The next chart is set for the date of King's "I Have a Dream" speech. Dr. King has no Kite pattern in his birth chart. However, on the day of the speech, transiting Neptune in the Mercury-centered chart completed the Kite pattern with his birth Flores, Jupiter, and Pluto. Transiting Pluto joined with transiting Neptune to complete another Kite pattern with Flores and Pluto. The birth placements indicate the power of Dr. King's speech. He influenced millions of people with his direct, heart-felt statements concerning the importance of civil rights in the United States. The "I Have a Dream" speech set the tone for the Civil Rights Movement, a movement that had the support of Americans of all creeds and colors.

With Flores integrated in the Kite, we immediately perceive the power of thought and word for Dr. King. His work centered on his ability to address people and deliver the truth so they could understand and adopt it. The involvement of Pluto in both the birth chart and the transit indicates the power that Dr. King brought to his message, and the power of the message itself. The

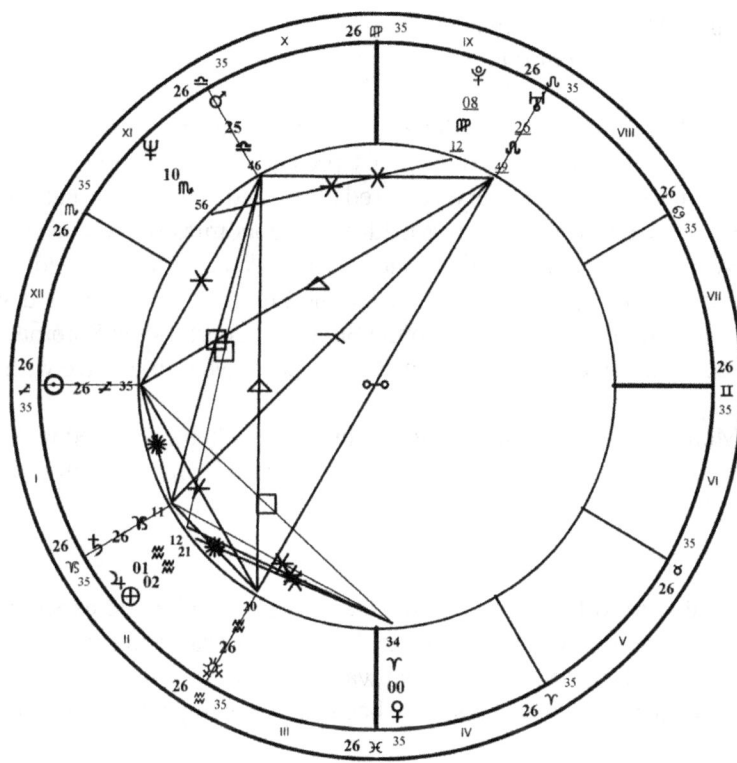

Mercury-Centered chart for Barack Obama
Birth Data: August 4, 1961, 7:24 pm AHS, Honolulu, Hawaii

fact that Neptune completes the Kites is highly significant. Neptune reflects devotion to a task, teacher, or cause. Dr. King lived his life devoted to the principal of equality for all Americans.

Barack Obama
Barack Obama was elected to the presidency 145 years after Lincoln's address, and 45 years after Dr. King's speech. Obama represents the power of the ideas spoken by both of his predecessors. He reached the position of president not through family power or through inheritance, but through his own effort. Politics aside, he embodies the spirit of Lincoln and the hopes of King.

Obama has no Kite configuration in his birth chart. We will have to allow history to determine whether he has comparable impact on the nation or the world, and then see if there is a Kite among the transits. In the Mercury-centered chart, however, there is a powerful configuration that includes Uranus, Mars, Sun, and Flores. Imagine, if you will, unfolding the pattern along the line of the opposition by moving the Sun from Sagittarius to Aries. This unfolding creates a

Mystical Rectangle. Therefore, within the trapezoid pattern in Obama's chart, we find the promise of the more mystical configuration.

It's really too soon to tell what may eventually turn out to be the most important words from President Obama. For the first time during his presidency, on February 8, 2009, the Sun transited to 24 Sagittarius in the Mercury-centered chart, and he faced the nation's deepest economic crisis since the Great Depression. World-wide economies weakened and/or failed, and Obama struggled to gain support for his economic plan. When Mars arrived at this degree in June 2009, Obama visited the Middle East. When the Sun reached the same degree in August 2009, Obama appointed the first Hispanic to the Supreme Court, Sonia Sotomayor. It's easy to see that the Mercury-centered chart maintains the expected Mercurial pace with its transits!

Examining the Mercury-centered charts of these three men when compared with each other, we find the theme carries all the way through, with Kites, Grand Trines, and other triangles of sextiles and trines reemphasizing the strong connections among them.

Charles Darwin
Being a close astral twin to Lincoln, Darwin shared many similar life experiences, including travel and writing. Darwin's work took years to develop, compared to Lincoln's speech written on the train going up to the ceremony. Darwin was no less constrained by the social system he lived in and he delayed publication of his work because of the impact he felt it might have on the Church.

Darwin has a Kite nearly identical to Lincoln's because they were born on the same day. Therefore we can expect this pattern to be prominent for important dates in his life.

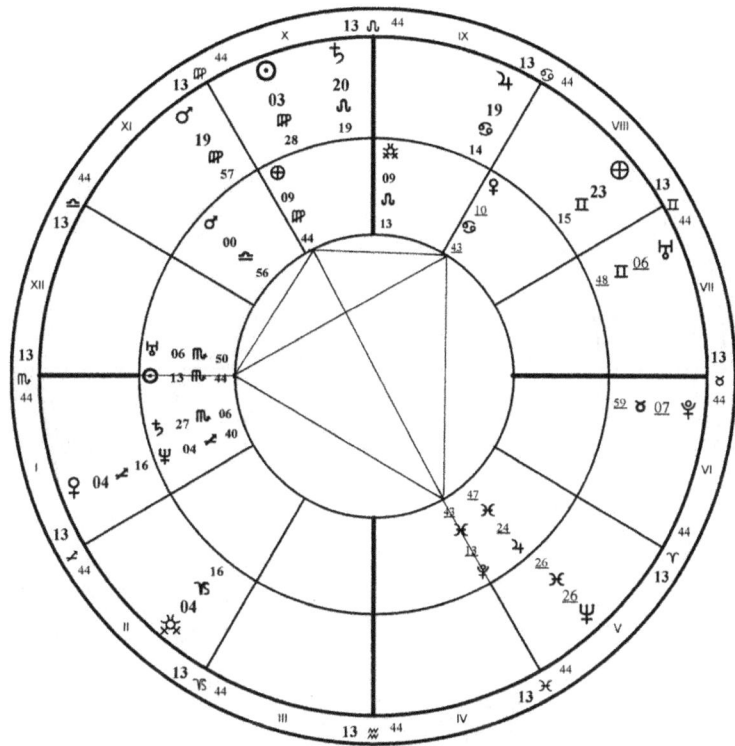

Mercury-Centered charts for Charles Darwin (inner) and Publication of Origin of Species
Birth data: February 12, 1809, 3:00 am LMT, Shrewsbury, England
Publication data: November 24, 1859, 12:00 pm GMT, Shrewsbury, England

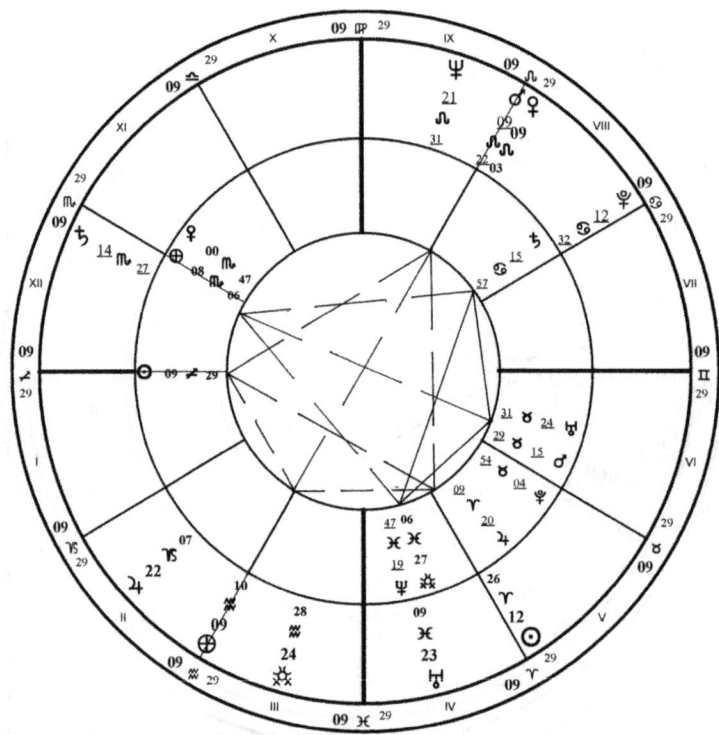

Mercury-Centered charts for Clarence Darrow and start of Scopes trial
Birth data: April 18, 1857, 7:50 pm LMT, Kinsman, Ohio
Trial start data: July 10, 1925, 9:00 pm EST, Kinsman, Ohio

Scopes Trial Charts

Charts for the lawyers involved in the Scopes trial share similar patterns as well! The Scopes trial provided a major milestone in the law concerning teaching evolution on American schools, an issue that survives to this day in parts of the country. Lawyers William Jennings Bryan and Clarence Darrow remain prominent figures in the history of jurisprudence, and the Scopes trial did nothing to diminish the stature of either man.

When considering the transits for the date the Scopes trial began, Darrow, the elder of the two lawyers, had a Kite in his Mercury-centered chart involving the Earth, Sun, and Venus/Mars. Additionally, a second Kite with wider orbs included Saturn, Neptune, Mars, and Pluto. This second Kite triggered a dynamic triangle in the birth Mercury-centered chart that included Neptune, Mars, and Saturn. The promise of communication, as seen in the Mercury-centered chart, included nearly hypnotic powers, diligent pursuit of the truth, and the capacity to structure arguments around complex emotional issues. Darrow came out of retirement to take the Scopes case for the defense, and the trial forms a significant part of his legal legacy.

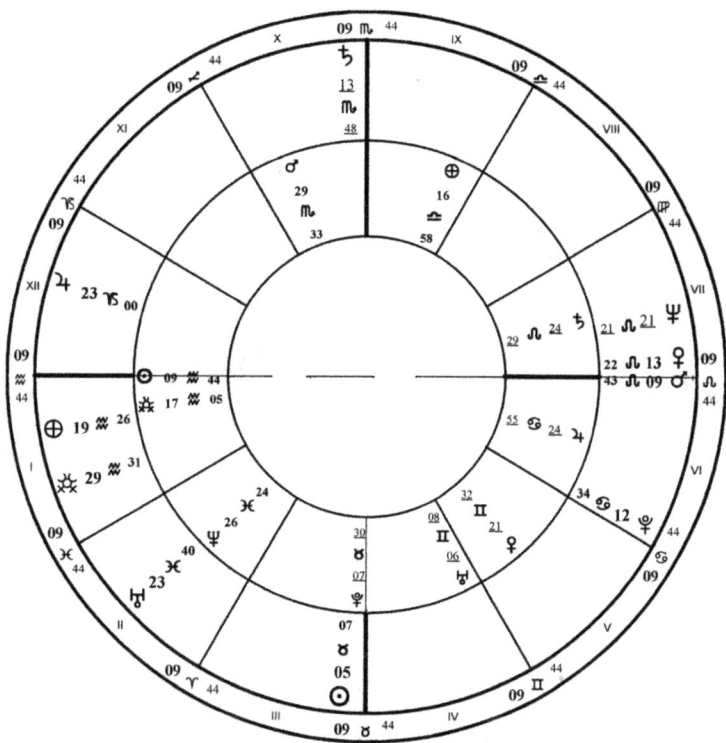

*Mercury-Centered charts for William Jennings Bryan (inner)
and his win of a key ruling in the Scopes trial (outer)
Birth data: March 19, 1860, 9:15 am, Salem, Illinois
Ruling data: July 17, 1925, 3:10:47 pm CST, Salem, Illinois*

Bryan was gifted with two Grand Trines in his Mercury-centered chart. On July 17, 1925, Bryan had Mars transiting to exactly oppose his own Sun. That day the judge ruled in Bryan's favor, saying that Darrow could not present expert testimony concerning evolution or its consistency with biblical scripture. This ruling caused Darrow to call Bryan as a witness, and the testimony demonstrated some of the emotionally-based beliefs at the crux of the case.

Regardless of that testimony, Scopes was found guilty on July 21, when Sun and Venus transited to square and oppose Flores in Bryan's chart. The ruling was later overturned on a technicality.

Repeal of the Butler Act
The Butler Act, under which Scopes was tried, was not repealed until May 13, 1967. It's interesting to note that Clarence Darrow "won" the trial, but it took 42 years to make it so.

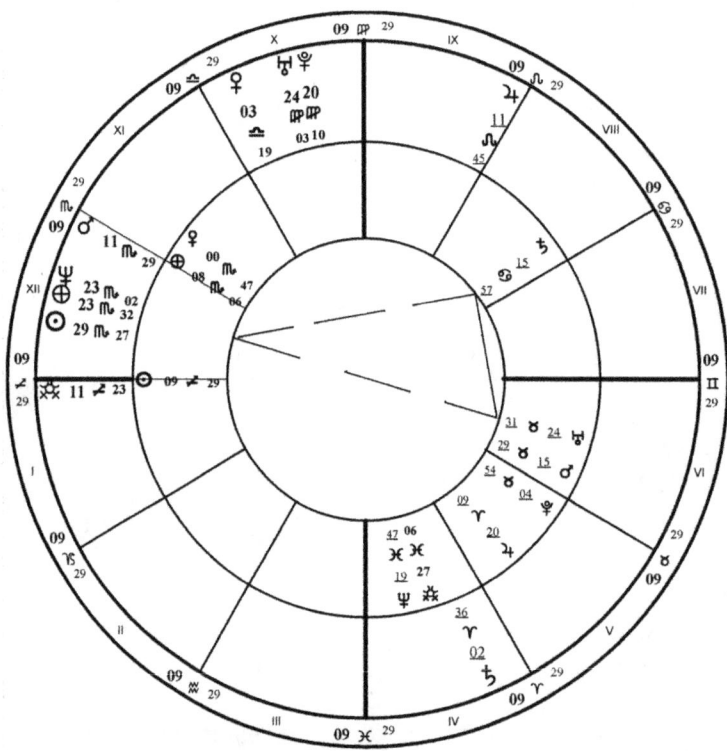

Mercury-Centered charts for Clarence Darrow (innr) and repeal of the Butler Act (outer)
Repeal data: May 13, 1967, 10:37:40 pm LMT, Kinsman, Ohio

These examples demonstrate the power of the Mercury-centered chart to reveal how our words and thoughts may produce lasting results. The Mercury-centered chart addresses our lives from a transpersonal perspective, and we observe that the charts may continue to reflect key issues long after the end of one's life.

On another note, in July 2007, Pope Benedict wrote about the concept of "theistic evolution," a theory he says comes from the "creative reason" of God:

> Currently, I see in Germany, but also in the United States, a somewhat fierce debate raging between so-called "creationism" and evolutionism, presented as though they were mutually exclusive alternatives: those who believe in the Creator would not be able to conceive of evolution, and those who instead support evolution would have to exclude God. This antithesis is absurd because, on the one hand, there are so many scientific proofs in favour of evolution which appears to be a reality we can see and which enriches our knowledge of life and being as such.

But on the other, the doctrine of evolution does not answer every query, especially the great philosophical question: where does everything come from?[8]

Summary

The Mercury-centered chart, with the inclusion of the companion asteroid Flores, suggests ways for you to pursue goals, improve intellectual processes, and mediate difficulties from the perspective of Mercury, the messenger. The nature of androgynous Mercury allows for any and all possibilities, unlimited by genetics or social status. Your personal situation has an influence, but for Mercury, all such boundaries exist largely to be crossed or overcome.

On the transpersonal level Mercury relates to the transformation seen in the sign Scorpio. Never underestimate the power of communication to be a change-agent in your life.

1. What elements do the Sun and Flores occupy in the Mercury-centered chart? How do these elements tend to interact?

2. What signs do the Sun and Flores occupy? What energy does Flores' sign offer to bolster the Sun's position?

3. Consider the sign the Earth occupies as that is where you are from Mercury's perspective. What do you learn about your own thinking process when you consider Earth's position?

4. How does the shape of the Mercury-centered chart differ from the geocentric chart? How does this difference reflect the perspective of Mercury? How can you use this difference in your own thought and communication?

5. Is there a distinctive chart pattern? How does that reveal Mercury's expression?

6. Are there any aspect patterns involving Flores? How do they amplify your understanding of how you might use the energy of Mercury?

7. Consider the sign Scorpio in the Mercury-centered chart. What does this tell you about the less obvious ways Mercury influences your life?

In the next chapter we will see how the Venus-centered chart sheds light on relationships of all kinds and clarifies issues of love and romance with the same precision we found about communication in Mercury-centered charts.

Chapter Three

Venus and the Venus-Centered Chart

In studying the mythological and astrological significance of the planet Venus, one becomes quickly aware of the ever-changing, ever-altered images of this archetypal energy. Throughout the historical development of human consciousness, the projections of the feminine onto goddesses have also changed and developed. Comparison of goddesses from different cultures reveals information concerning both the developmental process of feminine energy and cross-cultural similarities of feminine projections.

The mythological expression of Venus' energy has followed highly discriminated development. In Greek mythology, for example, Aphrodite began as the daughter of Zeus and Diana, a rather vague deity. "In origin Aphrodite was—like the great Asiatic goddesses—obviously a fertility goddess whose domain embraced all nature, vegetable and animal as well as human. Afterwards she became the goddess of love in its noblest aspect as well as its most degraded."[9] One myth has Aphrodite rising from the sea, the product of the meeting of Ouranus' blood with the waters ruled by Neptune. From either violent or natural beginnings, Aphrodite developed into a fully defined and refined expression of the best and the worst in the feminine.

In the Sumerian and Babylonian myths, Inana and Ishtar were goddesses of love and fertility who also had fearsome, warlike aspects. Some stories suggest they were the creators of wisdom, possibly through stealing it from Ea (Enki). Both descended into the underworld, also reminiscent of Aphrodite's marriage to Hephaistos, who lived and worked beneath a volcano.

To compare with another myth from the Egyptian culture, we find that Isis also began in humble and modest form as a protective deity. She became Osiris' wife and bore a son, Horus, and her popularity increased greatly after that.[10] The Egyptian Isis gains her higher position through marriage, apparently, but we also find that she has unusually well-developed talents along other lines.

Isis, indeed, was a potent magician and even the gods were not immune from her sorcery. It was told how, when she was still only a simple woman in the service of Ra, she persuaded the great God to confide to her his secret name. She did this through a magical torture that began with a charmed serpent which bit him. In his suffering he told Isis his name by another magical means—the name passed from his breast to hers. Only then did Isis conjure away the pain of the snake bites.[11] Here we find a seemingly simple lady who has hidden talents in abundance. This hidden talent, combined with her outer appearance, suggests the nature of the anima archetypal energy. Later she searched the world to find all the parts of her dead husband to bring him back from the underworld and death.

Isis used her magical powers, which had previously remained hidden; Aphrodite used her feminine wiles to seduce gods and mortals; Inanna used her ambitious drive for power to get what she wanted. All three display characteristics of the less unconscious anima as well as the more conscious feminine.

Another common element that has not been mentioned so far is the warrior-like quality of these goddesses. Isis sought far and wide for revenge of Osiris' death and she also developed funerary rites of embalming and mourning. Aphrodite tried to express herself in a war-like way but ended up being told by Zeus to go back to being her sweet seductive self, where she had more expertise. Inanna, through her newly gained knowledge, became capable of using weapons, the law, and even prostitution in her search for ultimate power. These goddesses were powerful, fierce, driving forces within their cultures; they developed greater strength as men projected more of their unconscious onto the goddesses.

These three expressions of the archetypal energy of Venus embody many facets of the planet's energy. For the modern astrologer, despite our removal in time and space, the goddesses are very much with us here and now on the collective level. The archetypal energy of Venus can be considered from four approaches: physical, emotional, mental, and transpersonal.

Venus in Astrology

Relationship is a principal expression of Venus. This planetary energy addresses the power we need to form relationships of all kinds. On the political, social, and personal levels, Venus guides our thoughts and feelings to help us to form lasting, valuable relationships.

In traditional astrology, a weakly aspected or positioned Venus indicates trouble in the area of relationships. At the transpersonal level, however, those same factors indicate ways to use Venus energies to help others with their difficulties.

Venus exemplifies the ultimate feminine force. She signifies an area of the life filled with the feminine—the intuitive, creative, and charming, the warrior-like, seductive, powerful—the Mother in all her forms. Let's look at four levels of expression of archetypal energy. Venus expresses on the physical level through sexual union. The beauty of the body is emphasized, as well as the beauty of the sexual act itself.

Venus Key Words

Physical	Emotional	Mental	Spiritual
kidneys	passion	choice	bliss
veins	sympathy	symbolism	harmony
attraction	calm	sublimation	love
gluttony	desire	refinement	wisdom
sexuality	interdependence	sociability	desire for Unity
			higher self of planet

Astronomy of the Venus-Centered Perspective

Venus is the most prominent object in our sky aside from the Sun and Moon. Sometimes called Earth's sister planet, Venus is slightly smaller than Earth. It's also our closest neighbor, approaching within 25 million miles (40 million km). From earth's perspective, Venus extends only 47 degrees from the Sun in either its evening star or morning star position. However, at this position, Venus is very bright and visible up to three hours after dusk or before dawn, depending on its position. The brightness of this planet in the morning and evening skies has stimulated rich mythologies around the world and astronomical observations abound in cultures in both the Eastern and Western Hemispheres.

Venus is close to the Earth in size and distance, yet the differences between these planets are profound.

First, conditions on Venus would not support human life. There is a lack of water. In addition, the atmosphere on Venus is 100 times as dense as on Earth. We simply could not breathe, even if the components of the atmosphere were suitable, which they are not. Venus, like Mars, has an atmosphere comprised mainly of carbon dioxide, while Earth's atmosphere is rich in nitrogen. The clouds on Venus are composed of sulfuric acid, producing an acid rain that evaporates before reaching the surface of the planet. The sky has a bright orange cast. One astronomy book likened conditions on Venus to traditional depictions of Hell. This observation is supported by the intense volcanic activity and frequent lightning bolts. Physical observations of Venus are the antithesis of mythological lore of Isis, Aphrodite, and other Venus-like goddesses.

There are almost no seasonal changes on Venus due to the fact that the inclination of the rotational axis is 177 degrees, or nearly vertical to its orbit. There is a runaway greenhouse effect on Venus because the dense atmosphere traps energy coming from the Sun, preventing the escape of heat into space. Surface temperatures on Venus are approximately 460 degrees Celsius.

Venus appears to have no intrinsic magnetic field. All the other planets generate their own internally driven magnetic field. This difference makes Venus a mystery in terms of cosmogenesis. The axial rotation of the planet is 243.1 days retrograde, causing the Sun to rise in the west and set in the east. The high altitude cloud rotational period, in contrast, is approximately four days. The noon-to-noon period on Venus is 116.8 Earth days, making one Venus year equal to 1.93 Venus days. Despite the slow rotation, the temperature on the dark side does not vary a great deal from the side facing the Sun.

For readers who want to immerse themselves in the lore of Venus, I recommend *Venus: Her Cycles, Symbols & Myths* by Anne Massey.[12] This book considers the geocentric perspective in great depth and also examines the transiting cycles of Venus, her phases, and her stations. Here I want to mention a few of the points that Anne elaborates so elegantly.

- The cycle of Venus conjunctions to the Sun and her stationary positions follow the golden mean, describing a nearly perfect five-pointed figure over a period of eight years. Thus the pentagram becomes a vital image associated with Venus.
- The number five has profound significance in the symbolism of various philosophies and religions.
- Venus is visible as a morning or evening star and is particularly striking when viewed in association with the crescent Moon.
- Venus is linked with fertility.
- Venus has phases similar to those of the Moon.
- The sign placement of Venus in the geocentric chart reflects how you "dress up" for the world in terms of appearance and thought.
- Venus retrograde is like being under a magnifying glass.

There is far more to be learned about Venus and her astronomy. A taste here helps you to understand the self-evident qualities of Venus that become the central focus in the Venus-centered chart.

Personal and Transpersonal Views of Venus

Dane Rudhyar said that Venus reflects interdependence. This is not codependence, but rather the interdependence all people have for others. Adults generally outgrow the total dependence of childhood and establish relationships based on mutual trust.

Although it is difficult to see ourselves in reference to others, there is a continuum on which relationships can be measured with respect to being more or less interdependent. The Venus continuum can also be described in terms of harmony-disharmony. This range is easier to objectify. A third continuum has to do with attraction and repulsion. Attraction can be defined in terms of appearance, language and body movement, to name a few possibilities, and the repulsion end of the continuum is verifiable in the same terms. The term "social" applies to Venus without obvious interdependence. Here the experience is one of interaction.

Venus-Centered Observations

Venus, in esoteric astrology, represents the Higher Mind—higher mathematics and philosophy. Its associated colors are indigo, blue, and bronze. Venus, the planet of Fifth Ray energy, relates to concrete knowledge developed through language and other social processes. Venus energy may be difficult to understand because it is the alter ego of the earth.

Venus relates to individual integration. Psychologically, integration includes concrete knowledge and the ability to process information. It also includes the capacity to integrate contents from unconscious recesses within the mind. Venus reflects intelligence as we know it. Intelligence emerges because we are both self-aware and aware of our environment and the people and things in it.

From the perspective of Venus at the time you were born, you can identify the most direct path for integrating your intellect and intuition. This integration leads to greater understanding of what the Universe has planned for you in the current incarnation. In other words, through understanding your Venus-centered chart, you gain understanding of your transcendent purpose.

The Sun represents the direction of personal mission and spiritual path. The Sun is the source of energy and life. It represents the Mind (Logos) of the Solar System in any chart. As such it is the most powerful point in the chart. The emphasis by sign describes the more physical attributes of life. The Sun's placement provides a focus for the conscious experience of Venus's energy.

In the Venus-centered chart, the Sun also provides the most powerful indication of your potential in relationships of all kinds. The Sun indicates the individual thrust of your energy as you engage with other people. The Sun's sign in the Venus-centered chart indicates:

- Coherent vitality of logic and reasoning—how you assimilate and use concrete knowledge.
- The role of magnetic attraction in your life—how relationships can work.
- The role of physical activity in your life—how you use your physical nature to express knowledge.

Venus principles to consider include love, art, physical attraction, sense of harmony, sexuality, sentimentality, and connection to center of art or entertainment. The down side of Venus includes harsh criticism, arrogance, and an opinionated posture. There may be blaming and a lack of expression of love. The upside of Venus incorporates a powerful, keen mind, mental control of emotions, and the ability to excel analytically. Qualities include truthfulness, fairness, independence, common sense, the ability to focus and concentrate, and unattached love.

If Earth's ego is a material and physical one, then "Venus is the Earth's alter ego."[13] The personal unconscious is the resting place for all the "hellish" things we do not like about ourselves. By contrast, "Venus is to the Earth what the higher Self is to man." Through the Venus-centered perspective, we gain profound understanding of spiritual matters.

Alchemically the tremendous heat on Venus suggests a continual "cooking" process. We might continually refine thoughts and feelings, stripping ourselves of limiting ego-centric material. From our physical perspective this would be an untenable situation. In the realm of the Higher Self, intense Light and heat may not be a problem.

Venus is fully empowered at two stages of human development: first at the point of individualization, when we experience differentiation of the ego and thereby realize the role of interdependence with other people; second, when we begin to realize our roles as beings of Purpose. Thus Venus is initially empowered when the ego develops, and again when we realize the powerful creative potential that we possess both as individuals and as members of the larger group.

The Venus-centered chart, then, reveals what we can understand about our relationship to the world in these three specialized contexts:

- Awareness and understanding of your desire nature. It is not so important to control these desires as it is to understand that they exist and that they are natural manifestations of the exoteric, physical realm of existence. Through pursuit of concrete knowledge of human desire, we move toward higher knowledge.
- Desire to resolve the opposites we experience in social situations and within our own minds. Material emerges from the personal and collective unconscious for consideration as we grow toward individual maturity. We work to see polarity not as an unresolvable paradox, but as a challenge—an invitation—to understand Unity.

- Spiritual mission to establish unity in place of divisiveness. We come to see that our truest goal is the attainment of informed Unity. We have fleeting experiences of Unity and we strive to understand even while we are in a physical, dual manifestation. In this way we cultivate the full power of informed process, developing our ability to manifest creatively in the world.

The Role of Juno in the Venus-Centered Chart

Juno, as the planetary companion of Venus, represents multiple individuals coming together or two individuals coming together multiple times. Partnerships are not only the joining of two people, but also the joining of the people they are connected to. The idea of coming together multiple times suggests that relationships have a long journey with many chapters to fulfill. This may occur within a single lifetime, or it may take place over a series of lifetimes.

Juno multiplies the expression of Venus energy. Instead of the individual integration within your own mind, Juno reflects the role of multiple integration—that is, integration through relationships. In the Venus-centered chart, Juno indicates where the individual seeks concrete knowledge through relationships. Juno focuses Venus' qualities in the direction of relationship. While Venus represents concrete knowledge and the capacity to think as well as the psychological tendency or demand for integration, Juno points to relationship as the key factor in integration. In many respects, human beings learn more about personal integration through the relationships than through any other single means.

Maritha Pottenger and Zipporah Pottenger Dobyns have delineated the influence of Juno in the relationship between mother and daughter with great clarity.[14] Their material translates effectively to other relationships if you set aside the nuances of the mother-daughter relationship. Juno can reflect powerful support and devotion. Juno also has the capacity to meet personal needs.

Juno reflects the capacity for self-knowledge and self-mastery, both of which lie at the core of Venus' potential. Juno shows how and where you may achieve balance in relationships, sharing with the partner and not leaning on him or her. Both partners then can achieve independence without forsaking each other. While modern psychology would counsel against codependent relationships, no healthy, lasting relationship can exist without interdependence. Juno shows one path, from Venus' own perspective, for achieving cooperation and support within relationships of all kinds.

Psychological intimacy produces deeply satisfying relationships. Many people experience sexual intimacy. However, psychological or spiritual intimacy emerges only when profound trust has been established. Juno in her role as wife and mother would encourage lasting relationships.

Thus the placement of Juno in the Venus-centered chart indicates how you may expect to develop lasting intimacy and trust in a relationship. Once that has occurred, spiritual ecstasy becomes possible.

Commitment, that ubiquitous term many people bandy about when they discuss long-term relationships, is at the heart of Juno's expression. Juno reflects how you may find the qualities that make for long-lasting romantic relationships (and other kinds of partnership as well), qualities like loyalty, fidelity, and mutual support. Make no mistake though. Juno does not demand that you roll over and play dead for your partner. Part of her charm is her cleverness and intellect. Juno will argue (hopefully fairly) to maintain her point, yet support her partner even when she does not completely agree. Thus Juno, as Venus' planetary companion, shows how you can develop qualities of loyalty and commitment to your partners.

Patterns in Venus-Centered Charts

Sometimes the overall pattern of the Venus-centered chart resembles the geocentric pattern in shape, but you find the planets in different arrangements within the pattern. Sometimes the Venus-centered pattern is radically different from the geocentric chart, placing emphasis on a different set of planets and aspects. In either case, the planets are still the same planets and the aspects are still the same aspects. The difference is the focus of energy of the center planet.

The Venus-centered chart reveals one way you can pursue relationships of all kinds, as seen from the perspective of Venus. The Earth's Moon is not present. Instead, Juno is part of the pattern. By examining the aspects and pattern of the chart, you can delineate and refine your understanding of relationships for the individual whose chart you examine.

Examples

Elizabeth Taylor
Elizabeth Taylor provides loads of research data in the realm of relationships. Here marital track record suggests that she was always searching for something in her love life. Her relationship with Michael Todd could have been "the one," and had he not died, she may not have had such a colorful range of partners.

While her Earth-centered chart provides plenty of information regarding relationships, her Venus centered chart focuses only on relationships and provides a profound perspective concerning her transpersonal partnership needs. The chart provided here includes only opposition, square, trine and sextile aspects. The patterns are stunning! No wonder relationships were so integral to Ms. Taylor's life.

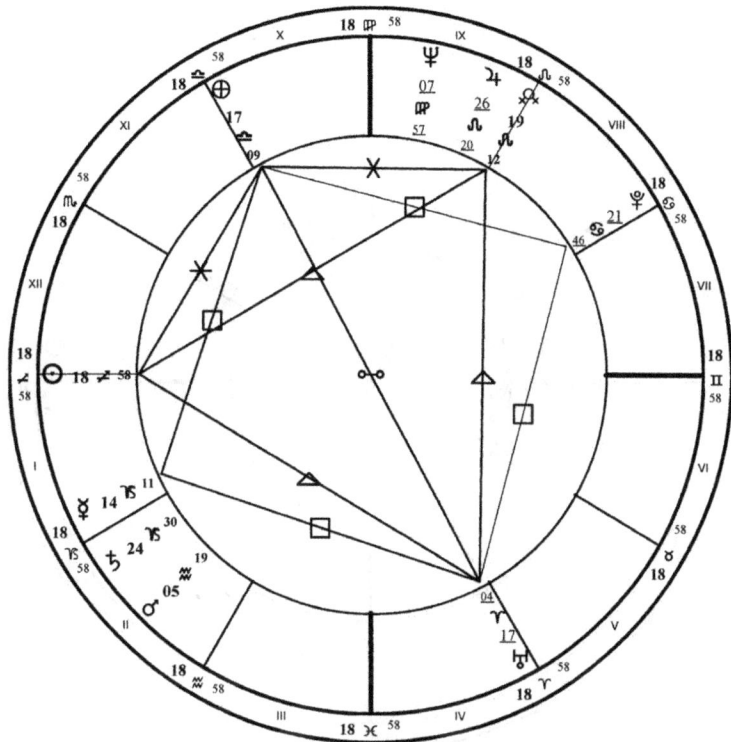

Venus-Centered chart for Elizabeth Taylor
Birth data: February 27, 1932, 2:00 am GMT, London, England

Juno, Venus's companion, participates with the Earth, Sun, and Uranus in a Kite pattern with tight orbs. The Grand Square has somewhat wider orbs, but we know that individuals grow into wider orbs through their lifetimes. The wide orb of the Mercury-Pluto opposition (not shown) can be expected to come into play at some point. Both the Grand Trine and Grand Square patterns focus on Uranus and the Earth in this chart.

The flow of romantic energy in Ms. Taylor's life followed distinctly different styles. The Kite with its Grand Trine suggests that she had a knack for easily identifying potential partners and drawing them into her life. Uranus suggests that intuition worked in her favor. Juno reflects the possibility for romance and marriage while Uranus indicates the propensity for ritual solemnization of romantic relationships.

The Grand Square, in contrast, indicates the challenging nature of partnerships and romance in Ms. Taylor's life. Here the power has shifted from Juno and the Sun to Pluto and Mercury. Her acting ability resided at least in part in her capacity to convince the audience of her feelings.

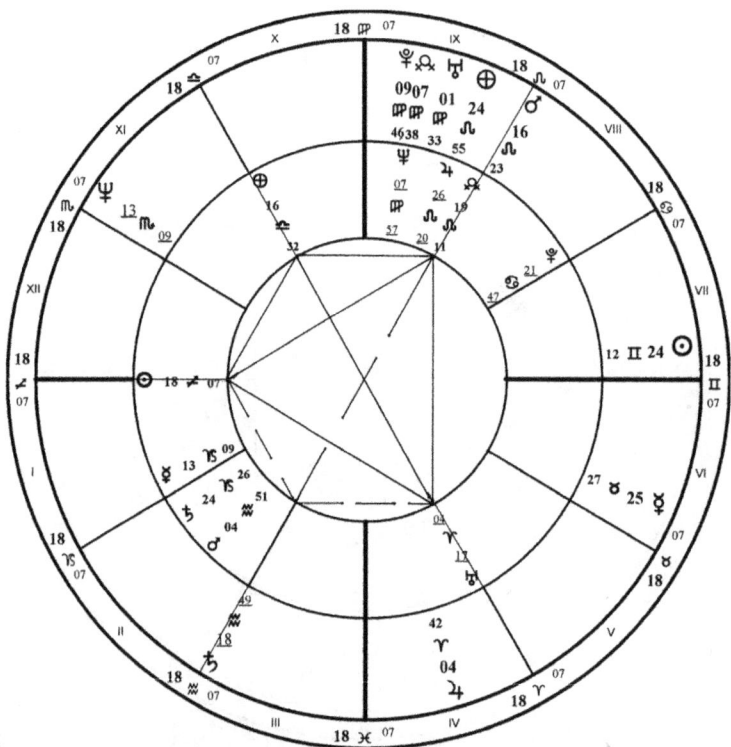

Venus-Centered chart for Johnny Cash (inner) and chart for recording of "Ring of Fire" (outer)
Birth data: February 26, 1932, 7:30 am CST, Kingsland, Arkansas;
Recording data: March 25, 1963, Kingsland, Arkansas

Pluto and Mercury in a wide aspect suggest less awareness of the impact of her words, with greater awareness of the impact of her actions as read in the Earth-Uranus opposition, the closest aspect in both the Grand Trine and Grand Square patterns.

Johnny Cash
Johnny Cash was born about twelve hours before Elizabeth Taylor. Although the orb of his Mercury-Pluto opposition is even wider, they share the Kite and Grand Square in their Venus-centered charts, so we can expect similar romantic paths in their lives.

On March 25, 1963, Johnny Cash first recorded "Ring of Fire," a song written by June Carter and originally recorded by her sister Anita. This song was about June falling in love with Johnny, and the song became his number one hit, recorded again and again throughout his career. The durability of the song is remarkable.

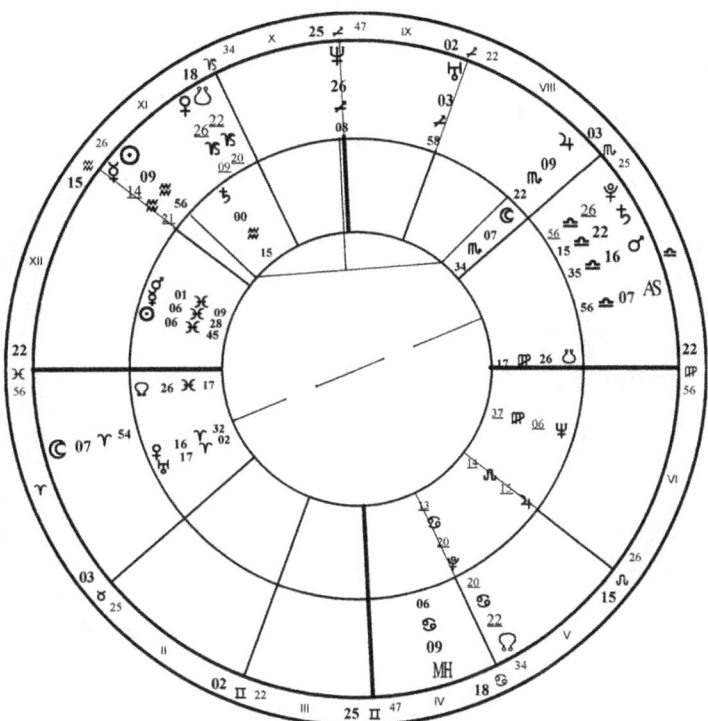

Venus-Centered charts for Johnny Cash (inner) and Adam Lambert (outer)
Adam Lambert birth data: January 29, 1982, 10:48 pm EST, Indianapolis, Indiana

In the Venus centered chart with transits for the recording date, Juno was conjunct Cash's Neptune and the Kite pattern was doubled by transiting Saturn in Aquarius, opposing Cash's birth Juno. The romance with June Carter lasted for the rest of Cash's life, reflecting Juno-Saturn perfectly.

Fast forward to January 29, 1982, the date Adam Lambert was born. The geocentric biwheel places Lambert's Neptune exactly in the midpoint of Cash's Sun and Jupiter at his (Cash's) Midheaven, indicating the mystical quality of good fortune in the association between the two. In addition, Lambert's Mars is *exactly* opposite Cash's Venus, highlighting an essential connection.

In the Venus-centered chart, Lambert's Juno and Saturn tie right into Cash's Grand Trine! So does Lambert's Sun if we increase the orbs allowed.

On March 18, 2009, Adam Lambert sang a middle-eastern version, complete with sitar music, on *American Idol*. The performance brought the house down! Adam took the song, sang a rendi-

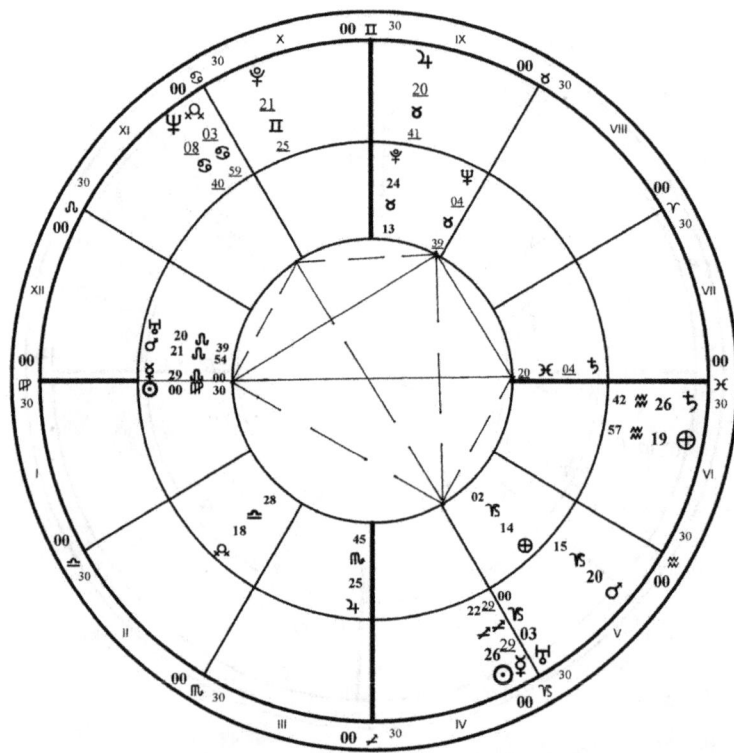

Venus-Centered chart comparison of Mata Hari (inner) and Greta Garbo (outer)
Mata Hari birth data: August 7, 1876, 1:00 pm LMT, Leeuwarden, Netherlands;
Greta Garbo birth data: September 18, 1905, 7:30 pm CET, Stockholm, Sweden

tion that he had heard, yet made it completely his own. Reviews, understandably, were mixed. Some viewers absolutely loved it, others were confused or found it distasteful compared to the Cash version. However, vast audiences were moved by Lambert's performance of a song written and recorded almost twenty years before he was born.

Fast forward again to March 18, 2009, the date of Lambert's performance. Transiting Juno was within 13 minutes of an opposition to Juno in Johnny Cash's birth chart, and transiting Jupiter was conjunct Lambert's Mercury within a degree. Frankly, astrology doesn't get much more self-evident than this.

Mata Hari
Mata Hari's name is synonymous with the spy trade, sexual intrigue, and the firing squad. In fact, her name made it into the *Oxford English Dictionary*, with the first listed literary reference fewer than 20 years after her death. Her Venus-centered chart includes powerful T-squares

composed of Jupiter opposing Pluto and squaring Mars and Uranus, a signature combination for romance and intrigue.

Greta Garbo played Mata Hari in a movie released December 31, 1931. Ms. Garbo's birth chart completes two Kite configurations with Mata Hari's chart. Garbo brings the energy of Uranus, Juno and Neptune to Mata Hari, fulfilling the promise of fame Mata Hari seemed so desperate to achieve.

Summary

Individuals change and grow in the crucible of relationships, and our interactions with others reveal our strengths and weaknesses on a daily basis. We acquire knowledge from these interactions, we face the power of sheer animal magnetism, and we learn about our unconscious depths as a result of both. Juno, Venus' asteroid companion, illuminates a second zodiacal point that resonates with Venus' energy at the deepest personal levels.

1. Venus in esoteric astrology relates to higher mind and also to psychological integration. How does the position of the Earth in your Venus-centered chart help you to understand your own higher mind? Where does your philosophical or spiritual knowledge take you?

2. What does the pattern of the Venus-centered chart suggest about how you can gather and work with knowledge?

3. What sign does Juno occupy? What does this suggest about the role of magnetic attraction in your life?

4. Is a particular aspect or aspect pattern strong in the Venus-centered chart? What does this say about which planets are more powerful from the Venus-centered perspective?

5. How does the Venus-centered chart reveal your creative potential, particularly in social groups?

6. Who or what do you naturally attract, according to your geocentric chart? Does the result get you what you want? How can you use the patterns and aspects in the Venus-centered chart to obtain better results?

In the next chapter we take our first look at a planet with moons. The planet-centered chart for Mars includes its moons Phobos and Deimos. Now we will see the first planetary system of moons that express multiple facets of the planet's energy.

Chapter Four

Mars and the Mars-Centered Chart

In one of the earliest comprehensive volumes concerning astrology, Ptolemy offers a definitive alchemical statement about the planet Mars: "Mars chiefly causes dryness, and is also strongly heating, by means of his own fiery nature, which is indicated by his color, and in consequence of his vicinity to the Sun; the sphere of which is immediately below him."[15] Based on the geocentric point of view of Ptolemy and medieval alchemists, Mars held a place in the heavens close to the Sun, and therefore directly participated in the Sun's heat and energy.

The nature of Mars throughout mythology has undergone significant changes. Mars in his role as the Greek god Ares was a terrifying god of war. He represented foolish courage to some, bloody rage and carnage to others. In the *Iliad* Zeus says of Ares: "Of all the gods who live on Olympus thou art the most odious to me; for thou enjoyest nothing but strife, war and battles. Thou hast the obstinate and unmanageable disposition of thy mother, Hera, whom I can scarcely control with my words."[16] For the Greeks, the argumentative qualities of the mother of gods were passed down to Mars in the form of masculine aggressiveness.

The Hindu god Karttikeya (also known as Skanda) provides a similar myth. Karttikeya was magically born out of the Ganges River and fulfilled his prophetic mission by killing the demon Taraka and restoring control of the universe to the gods. Often depicted riding a peacock and always carrying his bow and arrows, this god dresses in plain white with no decoration.

Mars, who began as a god of cattle and spring, held a place throughout Roman history as a god of agriculture as well as war. He was called Mars Gradivus. The term *gradivus* originally meant "to become big, or to grow," and later came to mean "to march." Mars as god of agriculture is totally consistent with this planet's rulership of Aries, the first sign of the tropical zodiac. The vernal equinox is the beginning of the zodiac as is also the beginning of the growing cycle for the northern hemisphere. Mars reflects the inspiration for this creative process.

Early mythological views of Mars give us a good starting place in our search for understanding of ourselves. Marc Edmund Jones stated in *The Guide to Horoscope Interpretation* that Mars clearly reflects human expression on the world, rather than within the personality. Mars is utilized by people to effect change in the physical world and indeed is the planet which represents physical energy. Mars deals with initiative and the practical daily operations of our lives.

In *The Astrology of Relationship*, Michael Meyer discusses Mars in terms of goal-oriented activities in the world. This externalizing of Mars energy carries the solar potential from within the psyche into the physical realm outside the personality. Meyer's statements focus on Aries/Mars without the confusion of Scorpio.

Karen Hamaker-Zondag has this to say:

> "The aggressive form in which the instinct for self-preservation occurs is symbolized by the planet Mars . . . Mars contents of the psyche can alienate an individual from his group. . . . In many instances the Mars function of man makes no direct contribution to inner refinement; its power of action, energy and self-assertion, ambition, aggression and pugnacity are pressed into service in the struggle against others."[17]

Here we see the preservative intention of Mars. If the expression of this intention seems destructive, it is because preservation of individual psyche and physical being may not always be consistent with the social requirements of others.

Culture dictates that individual qualities be balanced by social skills of sharing, harmonizing, and being considerate of others. Mars represents the tendency to resist social pressures and remain independent. Dane Rudhyar, in *New Mansions for New Men*, talks about Mars' quality of "moving away from." Organic life generates heat and this heat must be dissipated through activity as well as through the bodily processes; Mars functions as an outlet through movement and action.

From Ptolemy to Rudhyar another Mars theme emerges. Solar will cannot in itself manifest in actions. If the Sun shows what an individual can be, Mars shows how personal energy may pursue fulfillment of the Sun's will. Mars becomes destructive only when the connection between the Sun and Mars is stifled—that is, if Mars is not utilized as the initiator of action.

The table below shows how Mars operates on the physical, emotional, mental and spiritual levels. While Mars is action-oriented and tends toward expression outside the personality, it governs activities that emerge from every area within the psyche.

Mars Key Words

Physical	Mental	Emotional	Spiritual
body heat	determination	urge to act	devotion
muscles	self-discipline	impulsiveness	hara-kiri
sexuality	critical nature	quarrelsomeness	self-mastery
blood	fearlessness	passion	compassion
cerebral hemisphere	enjoyment of work	excitement anger	healing and healers
adrenalin	practical	enthusiasm	Salvation Army
burns	self-confidence		
accidents			

Mars energy represents a continuum of constructive and less constructive expression, with enthusiasm, determination and boldness on the constructive side, and defiance, jealousy, and violence on the less constructive side.

Because Mars is often thought to express primarily outside the personality, some information concerning the spiritual expression of Mars energy is important here. Alice Bailey, in *A Treatise on Seven Rays*, states that Mars governs the expression of devotion and that it has a spiritual effect on the discipleship or warrior level of being. The warrior in the Buddhist context is an individual who is on the spiritual path in every area of life, not just in the mental state. Such an individual has been through a crisis of initiation. Thus the physical nature of Mars is seen as not only beneficial, but essential to the disciple's process.[18] Many individuals seek direction and meaning in their lives beyond personal, self-absorbed levels of expression.

The energy of Mars clarifies the separation between opposites and thus serves a beneficial psychic function. Humanistic and transpersonal psychology point toward integration within the psyche. If we perceive no troublesome polarities in our lives, then we find no need for integration. It is only when we experience painful separation from Self or others that we realize change is required. Mars may stimulate awareness and start us on the path of integration.

Alexander Ruperti offers a fascinating discussion of the Mars-Sun cycle in his book, *Cycles of Becoming*. He begins with the idea that Mars energy is "almost always confused and egocentric." It will be confused to the extent that the personality does not have all the facts. If Mars energy is mainly unconscious, it will be painful and inconvenient.

The Sun-Mars opposition during Mars' retrograde period is the most important moment of the cycle. "The power which is released instinctively and without conscious awareness at the time of the conjunction can, at the opposition, be understood objectively and consciously in terms of the solar purpose determining its use."[19] When Mars is conjunct the Sun, it is closer to the Sun than to the Earth, so Mars directly combines its energy with the Sun. This is the point of strongest urge toward creation.

While the thrust of humanistic psychology is toward integration within the psyche, transpersonal psychology goes beyond, in the direction of integration of the body/mind with "other." The energy of Mars can aid both processes.

The period between a Sun-Mars conjunction and the opposition (about a year later) provides a natural rhythm of beginning and coming to awareness, allowing us time to assimilate the energies involved if we are aware of them. In the birth chart, the transiting Sun will conjunct and oppose the natal Mars once a year. However, the transiting Mars conjunction to the birth Sun may be more critical as it is the impact of Mars energy that brings awareness to the native.

The Mars transit reflects potential change within an individual. Mars indicates an outward moving energy. It shows the personal opportunity to go outside the psyche and to relate to the world as an individual.

Because Mars is outside the orbit of the Earth, the Mars-Sun opposition is the time when Mars is closest to the Earth, producing the hottest alchemical fire. There is time to review and revise the way Mars' energies have been used and to clarify the situation within the psyche. To the extent that we have developed respect for personal potential, we experience the Mars-Sun opposition as awareness of personal growth and progress.

Astronomy of the Mars-centered Chart

The Mars-centered chart has several new features, including its two moons Deimos and Phobos. These moons move very fast compared with the Earth's Moon: Phobos has a period of 7 hours, 39 minutes and 27 seconds, causing it to rise in the west and set in the east. Phobos is quite small, with diameters of 20 x 23 x 28 km. Deimos is even smaller, at 10 x 12 x 16 km. It is more than three times as far from the Martian surface, and has a period of 1 day 6 hours 21 minutes 15.7 seconds. Neither of these satellites is large enough to produce any kind of eclipse on the planet's surface.

In the Mars-centered chart the Earth will always be near the Sun, along with Mercury and Venus.

Due to the orbital speed of these moons (Phobos transits the zodiac more than three times per day and Deimos transits it in about 30 hours), their use requires exact timing of events and births. Phobos and Deimos are potential indicators in rectification of birth times as well as delineators of individual differences in multiple births.

Mythology of Phobos and Deimos

Phobos and Deimos have been characterized from antiquity by awesome mythological terror and fear. Terror, the sudden panic sometimes experienced, is associated with Phobos. Fear, a cultivated sense of the forbidding unknown (or the known), is given to Deimos. However, Phobos and Deimos also show the breadth of creative energy and provide hope for even the most difficult natal Mars.

Deimos and Phobos are the sons of Ares and Aphrodite. During the battle of Troy, Ares asked his sons to go among the soldiers, sowing terror and panic among them. The fact that their mother is Aphrodite provides an additional clue to their natures. Let's explore the difference between these two sons, keeping in mind that they will show us the nature of Mars' own perspective.

Traditional astrology imparts the quality of courage to Mars, yet his sons are the vehicle of terror and panic. The differentiation of fear into two parts refines our understanding of courage. Panic can arise when we least expect it. Soldiers can go through battle after battle and crack at the oddest times. Actors perform again and again, yet often experience stage fright.

Terror, on the other hand, is a thoughtful process. It arises as the result of rumination about what is to come. We can work up remarkable anxiety if we dwell on the problems we face. A soldier may be in a complete state of terror and still fulfill his or her role in the battle plan. The soldier is no more or less courageous than the individual who panics. He or she is simply equipped differently on the emotional level.

Ares and Aphrodite provide complementary energies to both Phobos and Deimos. Ares shows the aggressive side of Mars energy, while Aphrodite is filled with beauty and social graces. She embodies love. Esoteric astrology speaks of the devotion of Mars, a Sixth Ray planet. Aphrodite embodies devotion or its object in traditional astrology.

Phobos focuses on action, while Deimos focuses on thought. Phobos is compulsive, while Deimos is obsessive. Passion has its active and thoughtful qualities, so Mars is active and thoughtful in all applications.

With Aphrodite as their mother, both Phobos and Deimos incorporate elements of harmony and balance. Aphrodite (Venus) suggests attraction and love. The thought processes that precede artistic, creative expressions often include considerations of refinement and balance. The esoteric expression of Mars is devotion; the exoteric warlike side is not the only expression.

Esoteric astrologers suggest that anger and violence are weaknesses of the Sixth Ray, while love, imagination and visualization are strengths. However, these qualities are all facets of one energy. When idealism becomes fanaticism, then Sixth Ray energy is twisted; when intuition and sensitivity lead to greater understanding and respect for others, then the Sixth Ray expression of Mars is more direct.

Scorpio, the second sign ruled by Mars, represents the season in which life is terminated. Seeds are saved at this time for spring planting, and everything but the most hardy becomes dormant or dies. Astrologers have come to think of Scorpio as ruling birth and death. As far as Deimos is concerned, it is, rather, death that brings us to a new birth. Deimos deals with the energy of inspiration and the seed of thought.

It is important to note that Phobos and Deimos do *not* represent opposite sides of Mars' nature. In the same way that we tend to polarize the nature of the Sun and Moon into opposites, we want to polarize these two moons, but this would be an error. The relationship is more like ebbing and flowing waves. The water is the same; only its direction has changed, and even the direction is not precisely opposite.

Dane Rudhyar, in *New Mansions for New Men*, states that compassion is an expression of desire in its purest form. The will to wholeness within each of us contains both the active desire of Phobos and the contemplative compassion of Deimos, two sides of the full expression of Mars.

Deimos and Phobos, as aspects of Mars' energy, resolve an apparent polarity. On the one hand we have fully empowered independence as an archetypal principal. On the other hand we have a devotion to "other," whether that other is a person, a god, or an ideal.

Our experience of Mars in the astrological chart has many facets; Deimos and Phobos, through their own qualities, provide a richer picture of the potential of Mars. By examining their positions in Mars-centered birth charts of individuals, we can identify areas of life energized by Mars. We can also consider these degree placements in geocentric charts as areas of potential sensitivity and energy, each governed by the nature of the Martian moons themselves.

Considering the Aries/Scorpio rulership of Mars and Ares/Aphrodite parents of Phobos and Deimos, the following characteristics emerge:

Phobos Expression	*Deimos Expression*
Acceleration	desire
Accidents	inspiration
Agitation	emotions (generally)
Boldness	devotion
haste	fascination
creative action	creative thought
the passion of action	the passion of thought
compulsive action	obsessive thought

Utilizing a broad, nonjudgmental interpretation of Mars energy, Phobos and Deimos can show the mind of Mars in people's lives. Where does the mind of Mars take us? The answers to this question reveal how to flow with Mars instead of feeling controlled by it. Further, how can we cultivate devotion in my life instead of aggression? How can our physical activity draw upon compassion instead of on anger? We can resolve these questions by examining the Mars-centered chart.

Phobos Deimos Combinations

Because Mars' Moons, particularly Phobos, speed through the chart, the angular relationships between them change very quickly. Phobos transits the signs three times in only one day, resulting in a full range of combinations in fewer than eight hours. Using a two-degree orb applying and separating for semi-square, for example, such an aspect exists for approximately six minutes. Even twins can have these moons in significantly different angular relationships. At the very least, one twin might have a close aspect while the other does not.

The action/thought relationship of the moons may be used as a key to individual differences relating to thinking style, decision-making processes and overall approach to life. Research has shown that when Phobos rises ahead of Deimos, action precedes careful thought. When Deimos rises first, the individual has greater capacity for thoughtful planning, even in highly stressful situations. Aspects reveal a lot about this action versus thought personality trait.

Angular Relationships Between Phobos and Deimos

Conjunction (both in same sign): Desire and action will tend to follow closely upon each other with the fire of creativity at the heart of action. Passions will include thought and action at once. Mars activity is unified here.

Semi-sextile (in adjacent signs): The individual will experience thought and action separately. If Deimos is in the earlier sign, then thought will tend to precede action, though by a rather short period of time. If, however, Phobos is in the earlier sign, the individual always seems to act first and think later. This individual will need to tame the urge to act in order to develop a creative process that bears significant fruit. Too quick to act indicates boldness to the point of recklessness.

Semi-square and Sesqui-square: Action appears to result from an internalized motive or source, as compared with typical outward expression. The individual tends to contain his or her energy, appearing to be far calmer than is actually the case. Much of the individual's activity will take place on the inner plane. Other people may not perceive the difference between thought and action in the individual's behavior.

Sextile (signs of the same polarity): The moons of Mars will work in a harmonious relationship that maximizes opportunities for the individual. Action and thought flow smoothly in both directions, each creating opportunities for the other. Such an individual will have skill in deciding how and when to act, an intuitive sense of timing.

Square (signs of the same quality): Here we find very high energy. Deimos and Phobos in incompatible elements activate contrast. Challenging differences between thought and action sometimes leave the individual out of control. The aggressive warrior Phobos lacks patience to plan ahead as the farmer must. When rising first, Deimos can stall the person with obsessive thought patterns.

Trine (signs of the same element): Thought and action will have the appearance of being one. When in fire, the Mars energy will be channeled to creative thought as well as action. The individual will be able to manifest the creative energy in the physical realm. Moons in earth signs indicate solid practical thought and action. Passions will tend toward material expression. Deimos and Phobos in air signs will be the most objective. Mars energy will be utilized in an abstract way, or perhaps indirectly. Water sign moons will be emotionally based, depending on an inner sense of what is right or proper, possibly resulting in the most rhythmic relationship between thought and action. Even though thoughts and action are quite different, they tend to support each other naturally.

Quintile and Bi-quintile: If Phobos rises first, creativity depends on the capacity for action. When Deimos rises first, much thought goes into a plan before the first line is drawn or the first word is written.

Quincunx (five signs apart): While the moons are of incompatible elements, they are visible to each other and therefore capable of creative adjustment between action and thought. Such adjustments can involve high drama.

Opposition: Deimos and Phobos will have the greatest awareness of each other in this relationship. Boldness will be supported by a studied courage; the urge to independent creative process will be balanced by the desire for support within a group. The result: remarkably mature judgment.

Mars-Centered Examples

John Elway/Denver Broncos

Mars is the planet of energy and devotion to a task, person, or belief. John Elway, quarterback for the Denver Broncos, exemplified the exertion of physical energy throughout his career. He also exemplified the passion and devotion to the sport that made him a truly great player. He established a foundation to prevent child abuse and to help abused children, another example of his devotion to his beliefs.

Elway was drafted by Baltimore in 1983, but he refused to play for them. They traded him to Denver and he received a six-year contract worth more than $12,000,000.00. Therefore it makes sense to look at the Denver Broncos and Elway charts together.

The chart for the Broncos has a Bundle pattern. Within the bundle is a triangle including Saturn, Pluto and Neptune. Deimos, Mars' companion, aspects all three planets within a degree and a half or less.

Elway was born only a year after the Bronco franchise was created. When his geocentric birth chart is combined with the Bronco chart, a double Kite pattern is formed. A speculative birth time of 11:30 would place Deimos at the open point of a Grand Sextile pattern. The two charts together make a nearly perfect fit that reflects Elway's tremendously successful career in Denver.

Considering the first and last Super Bowl games Elway played in, where the Broncos lost the first and won the last, the following aspects occurred during the first two hours of each game (note: if you use different orbs, you will get different results):

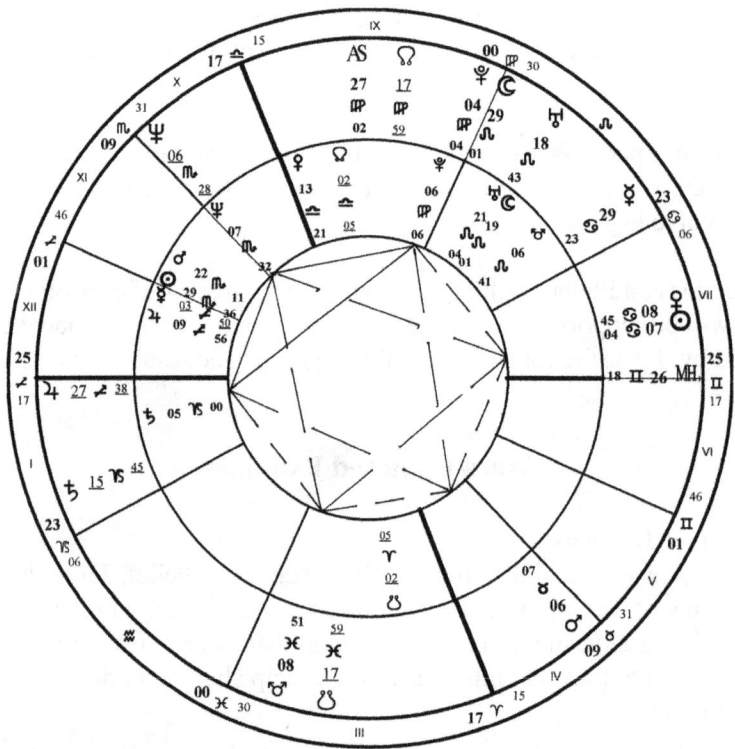

*Geocentric Comparison of Broncos team chart (inner) and John Elway (outer)
Broncos birth data: November 22, 1959, 9:00 am MST, Denver, Colorado;
John Elway birth data: June 28, 1960, 11:30 am PST, Port Angeles, Washington*

- Super Bowl 1987: 6 quincunxes, four squares, 2 oppositions, 1 trine and 1 sextile
- Super Bowl 1999: 5 semisextiles, 4 sextiles, 4 trines, 4 squares, 1 conjunction.

These numbers confirm what you might expect: a person like Elway who has many trines, sextiles, and semisextiles thrives under conditions where these are the transiting aspects. These results show that so-called soft aspects are activated by transiting soft aspects.

The Broncos First Superbowl Win
The Broncos' first Super Bowl win in 1998 reveals the power of planet-centered astrology. A biwheel from the Mars-centered perspective reveals the story. At 3:00 pm Denver time on game day, the chart was lit up by both a Grand Square and Grand Trine, promising all the action you could want plus a hefty dose of good luck. By the start of the game (3:24 pm PST), transiting Phobos had moved 19 degrees to square Phobos in the Bronco's chart, with transiting Uranus opposing the Bronco's Deimos. Was this the game to change the Bronco karma and deliver a win?

Mars-Centered Super Bowl Transit to Bronco Team Chart

Green Bay scored first on fewer than a dozen plays, but then, after about 22 minutes of clock time, transiting Phobos formed a conjunction with the Broncos' Venus in the first house. The Broncos also scored on their first possession, and the game was on!

Because Phobos and Deimos move so fast, a number of aspects were formed as the game went on. With two minutes left in the game, the score was tied. At 6:46 pm, transiting Phobos formed a conjunction with the Broncos' Neptune. Deimos was trine both natal Neptune and natal Venus. The Broncos scored and won their first Super Bowl victory. What really happened? The Green Bay coach decided to let the Broncos score! He felt certain they would score anyway and he wanted to preserve time on the clock for a quick run at tying the game. No such luck. The game ended at 6:49 pm with transiting Deimos still within orb of the trine to Neptune!

Sports fans are always looking for ways to predict wins. In the case of football games, you often will not be able to guess the actual starting time. Yet the chart for the starting time tells the tale in many cases. For those who wish to test the planet-centered approach, the 1999 Super Bowl began at 6:25 pm EST in Miami, and the game ended at 9:43 pm.

Summary

Already we begin the see a clustering of planets around the Sun as we move to the outer planets. Venus, for example, can only be about thirty degrees from the Sun in the Mars-centered chart. Here, the Earth reaches about forty degrees distance from the Sun. So from Mars' perspective, the mental flexibility of Mercury (capacity to organize knowledge), the attraction of Venus (power to attract things to ourselves), and the effective action associated with our lives on Earth are closely related to the Solar mission, at least from Mars' perspective.

1. How do the positions of Deimos and Phobos reflect your understanding of how you naturally tend to use Mars energy? How can you use this knowledge to help your use of active energy each day?

2. How do the elements/signs of Deimos and Phobos interact with each other?

3. Is there an aspect between Deimos and Phobos? If yes, consider the nature of their interaction. What does this tell you about how you tend to behave when angry? When enthusiastic? If no, consider what you can do to draw them together to help you? Is there another planet (or Earth) in aspect to both of them? How can that planet be used to get the two sides of Mars together?

4. How do you distinguish between the Deimos and Phobos expressions of Mars in your life? Think of a time when you were in full Phobos Mode. Then think of a time when Deimos 'energy dominated. How does this help you understand ways to use the energy of Mars?

5. How do Deimos and Phobos help you understand the balance between independence and devotion to others in your life?

6. The sign/element of the Earth in the chart shows where you are from Mars' perspective. What does this "energy" picture suggest about events that arise suddenly in your life? How does Mars tend to respond? Is that how you respond? You might practice taking action of the sort indicated by the Earth in the Mars-centered chart to see how it feels.

7. Remember, the birth time becomes a huge factor in planet-centered charts because of the speed of the moons. Phobos moves more that a sign and a half in one hour. If your birth time is off by even ten minutes, Phobos can change position as much as Earth's Moon moves in a whole day!

The speed of Phobos and Deimos underscore the active energy we associate with Mars. They also reveal two equal partners—contemplation (consideration) and action—in our efforts to fulfill desires.

The following chapter on Jupiter takes us into a new realm of complex relationships among satellites and how they remind us of the many ways we can acquire and use wisdom to support our spiritual lives as well as our material goals.

Chapter Five

Jupiter and the Jupiter-Centered Chart

Jupiter was the son of Saturn, father of Mars, and therefore grandfather of Romulus and Remus, founders of Rome. Jupiter was the Roman god most equivalent to the Greek Zeus. His earliest associations with the elements, particularly storms, later expanded to include the role of protector of the Roman people. His name is compounded with second names in Latin to indicate the particular function or need people were praying to him about. For example, Jupiter Liber was a creative force, while Jupiter terminus dealt with boundaries. Jupiter Optimus Maximus was known as the great god of the Roman Empire.

Zeus was the Greek god who ruled on Mount Olympus. In the *Iliad*, Zeus sent storms to plague his enemies. Transitioning from the protector of kings and their families, Zeus became a judge and peace-maker. In mythology he was the deity who settled squabbles among the other gods.

After his siblings had been devoured by his father Cronus, Zeus was hidden from his father and raised in secrecy. Eventually he overthrew Cronus and freed his siblings, Hades and Poseidon. The three then divided rulership of the universe, setting aside Mount Olympus as common ground.

Zeus, like his Roman counterpart, had many lovers. Among them were Metis, mother of Athena; Hera, mother of Ares, Hephaestus, and Hebe; Leto, mother of Apollo and Artemis; the mortal Leda, mother of the twins Castor and Polydeuces, as well as twins Clytemnestra and

Helen of Troy; Danae, mother of Perseus; Europa, mother of Minos, Rhadamanthys, and Sarpedon; Themis, mother of the Horae (goddesses of the seasons) and Moirae (the Fates); Mnemosyne, mother of the nine muses; the list goes on. Zeus's children each supplied something that people needed to be happy and successful.

In many cases we can track deities from other traditions through associations with Greek gods. Zeus himself was a tribal deity of thunderstorms. Tyr or Tiwaz was an early sky god of the Germanic people. Dyaus held a similar position in the Vedic pantheon. The names of all these gods come from root words that mean bright or shining.

In SELF-EVIDENT ASTROLOGY™, Jupiter is fittingly associated with family, given that he had many lovers and sired many children among the gods and humanity. The Galilean moons of Jupiter were all named after Jupiter's consorts.

Jupiter is the ruler of function and process. The form of things is rigid and relatively changeless. Function, on the other hand, has a dynamic changing quality. Even when the form of an object suggests a specific label, we are not limited to one specialized function just because of the label. A screw driver can be used to dig a hole and a hat can be used to carry water. Through the planet Jupiter we can investigate the function of everything in a chart, evaluating the success or failure of each planet's processes.

Ruling expansion, Jupiter, in mind and body, demonstrates its functional urge minute by minute through the diastolic function that expands the heart as it fills with blood. The growth process of the physical body involves expansion and division of cells as well as swelling of inflamed areas or growth of abnormal tissue.

Jupiter reflects the function of digestion and assimilation. Jupiter rules Vitamin B-6, essential to the absorption of the other B vitamins.[20] Without this basic nutrient the chain of assimilation is broken and body function is reduced. Biotin and choline are also ruled by Jupiter. These two nutrients are needed for proper metabolism of fats and proteins. Jupiter is also associated with the liver, blood, and function of the organs in general. The relationship between Jupiter and other chart factors will determine the personal nutritional and dietary needs of each individual. A natural remedy for physical difficulties is a diet that allows expression of Jupiter's wisdom at the physical level.

As we grow and learn, we experience the expansion of mental and emotional capabilities. The physical craving for particular nutrients is paralleled by potential emotional desires that can lead to excess. The mind can become greedy, expanding until it cannot be stretched any more. The limitations of form dominate, causing contraction to a more contained posture. Emotional expansion comes through the desire to have more and livelier experiences.

We seek wisdom, another Jupiter quality, as well. The spiritual expression of Jupiter is intuitive. When we pursue spiritual mission, we are asking Jupiterian, process-oriented questions. Wisdom and love are the result of such questions and their answers, if we are capable of understanding the insights offered to us.

The Alchemy of Jupiter

The alchemical process of purification, in its broadest context, is the Jupiterian heart of the work. The alchemist proceeds by stages in which the *prima materia* is somewhat purified, through dissolving, for example, and then resolidified. One observes the effect of the process and selects another process for further work.

Words such as separate, digest, and distill apply to alchemical processes that are easily understood in terms of daily life. Other alchemical terms such as fixation, inhibition, and projection relate to psychological processes. The alchemist uses all these and other terms to describe separation and unification, dissolving and making solid again, creating movement or establishing stillness.

The conscious mind employs logical, rational processes that can be communicated clearly to others. The unconscious mind employs irrational, circular processes that are far more individual and difficult to share. Jupiter allows us to experience both conscious linear thought and the emergence of formerly unconscious and irrational ideas into awareness.

Jupiter instills us with faith. We need to have faith that our individual journeys have destinations and that we will gain something through the journey.

Within the psychological journey that includes emotional, mental, and spiritual facets, Jupiter reflects forgiveness. Saturn gives us the structure of the Law but Jupiter gives us the spirit. If we must observe the letter of the Law, we are surely lost, but if we practice the spirit of the Law, then we fulfill God's will. Psychologically, too, if we must be "perfect" models of human beingness, we are lost. However, if we accept our limitations and foibles, then we can progress toward spiritual goals. Jupiter places emphasis on the functions of being human, rather than on the form of human perfection, whatever that may be. Experiencing the full range of possibilities on the physical, emotional, mental and spiritual levels will result in the most constructive and holistic functioning of the body/mind.

To consider the function of Jupiter without considering the form of Saturn is futile. The function of a transistor is meaningless without the form of the mechanism within which it can operate; the function of our blood is meaningless outside the body. The function of physical existence becomes absurd if we feel the need to be outside the physical form.

Jupiter Key Words

Physical	*Emotional*	*Mental*	*Spiritual*
blood	optimism	social sense	religion
nutrition	expansion	philosophy	aspiration
liver	joviality	judgment	prayer
veins		knowledge	prophets
			mercy

The Jupiter-Centered Chart: Focus on Process

Function—or process—is a dynamic part of life. The Jupiter-centered chart provides an expanded perspective on process in individual lives. Jupiter emphasizes change. Jupiter and its satellites provide a transpersonal model to compare with the geocentric personality model. Jupiter's placement by sign in the geocentric chart shows the focus of process; the Jupiter-centered chart develops choices that are keyed to the individual personality's desires and capabilities.

Physical Function
The Jupiter-centered chart amplifies the geocentric perspective by detailing the style of individual process. For example, suppose a person has Jupiter in Virgo, suggesting an earthy, analytical style. The Jupiter-centered chart has only Uranus in earth, with a concentration of Jupiter's moons in air and fire. We know that the effective practical style of Jupiter in Virgo depends on a creative, mental, and intuitive awareness.

Emotional Process
Jupiter's position also indicates how emotional processes work. For example, its placement indicates how emotions are triggered. Aspects to Jupiter indicate which planetary energies contribute to emotional processes. The Jupiter-centered chart provides possibilities for more creative uses of emotional patterns. It offers suggestions based on the way Jupiter's Logos is keyed to the individual personality.

Mental Processes
People tend to have two types of mental process: one is the logical, rational, linear thought of deductive reasoning; the other is the circular, non-rational, somewhat less conscious inductive process. Natal geocentric aspects to Jupiter show how mental processes move for an individual.

Spiritual Processes

Finally, Jupiter indicates spiritual processes. Development of the spirit depends on intellect, strong judgment grounded in feeling, some level of intuitive mastery, and clear perception. The moons of Jupiter delineate these processes via their placement and aspects in the Jupiter-centered chart. Moons in prominent patterns provides a wealth of information about the dynamics and outcomes of spiritual development.

Planetary Roles in the Jupiter-Centered Chart

Sun: the individuality of Jupiter, Jupiter's love wisdom and mission.

Mercury: communication, harmony/conflict, common sense.

Venus: sociability, concrete knowledge, love.

Earth: humanness, intelligent activity, the role of personality.

Mars: independence, devotion, creative activity.

Saturn: structure, intelligent activity, patience.

Uranus: intuition, ceremonial magic, equilibrium.

Neptune: prophecy, compassion, strong imagination, idealism.

Pluto: power, will, directed mind.

Major Satellites of Jupiter

Although Jupiter has many moons, the present discussion will consider the four largest: Ganymede, Callisto, Io, and Europa. The four largest were all discovered by Galileo in 1610 AD.

Ganymede is larger than the planet Mercury, with a diameter of 5262 km. It is the largest planetary satellite, has low density, and no atmosphere. Its synodic period is 7d 3h 59m 35.9s. This moon is completely inert at this time, with a few relatively smooth craters. Ganymede has an icy crust covering water that may be as much as 800 kilometers deep. Yet there seem to be no effects from tidal stresses.

The next is size in Callisto, with a diameter of 4800 km. Callisto's period is 16d 18h 5m 6.9s. In synchronous orbit (the same side is always facing the planet), it has a heavily cratered surface and no volcanic activity. Callisto also has water beneath a surface of ice. This is the only Galilean moon positioned outside Jupiter's radiation zone.

Io has a 3630 km diameter (slightly larger than earth's Moon), with a period of 1d 18h 28m 35.9s. This moon has temperature ranges from a relatively warm 17 degrees Celsius to -146. It receives high radiation from Jupiter. Io has active volcanoes. This activity may be related to tidal stresses as the surface of the moon may be seas of sulfur just below a very thin outer crust.

Europa is the smallest of the Galilean satellites of Jupiter, at 3138 km diameter. There are no volcanoes or craters. Europa has an incredibly smooth surface, probably ice. There may be huge quantities of water beneath the icy surface.

The periods of these four moons vary considerably. Io is fastest, at 1.75 days. Europa is approximately double that, with Ganymede about four times, and Callisto more than eight times the period of Io. The approximate ratio of 1::2::4::8 results from the combination of each moon's mass with its distance from the planet.

Mythology of Jupiter-Centered Moons
The mythology of the god Jupiter (or Zeus) is rich with examples of his passion, and each of the four satellites bears the name of one of Jupiter's conquests. Io, a mortal, had a dream that told her to let Zeus embrace her. Her father, upon hearing this dream, consulted an oracle and then approved the relationship. Hera sent a gadfly to torment Io, who ultimately found peace on the banks of the Nile River. Zeus restored her human form; there she raised her son. At times Io has been confused with Isis and her son with Apis.

Europa, the daughter of Agenor and the sister of Cadmus, was abducted by Zeus, who took the form of a bull and carried her off across the sea from Phoenicia to Crete. Europa there became the mother of three sons, Minos, Rhadamanthys and Sarpedon, who became judges of the underworld. Later, Europa married the King of Crete.

Ganymede is the only male in this group of moons. He was the son of Tros, King of the Trojans, and was considered to be the most beautiful of mortals. He was carried off by Zeus in the form of an eagle to become cupbearer for the gods. Because Tros mourned the loss of his son, Zeus gave him two divine horses (and also a golden vine) as compensation. Ultimately Ganymede was granted immortality, and placed among the stars in the constellation of Aquarius.

Callisto was either a nymph or the daughter of Lycaon or Nycteus. Sources generally agree that she was a companion of Artemis and that because of that role she had taken a vow of chastity. Zeus took the form of Artemis and raped her. He then turned her into a bear to hide her from Hera. Callisto was later shot by Artemis's arrow, possibly due to anger over Callisto's lost virginity.

Zeus placed Callisto and her son Arcas in the sky as the Great Bear and Small Bear constellations. This angered Hera and she convinced Neptune to refuse to let those constellations fall into the sea, so they never set. There remains some confusion in the mythology between Callisto and Artemis, who sometimes took the form of the she-bear.

Alchemical Expressions of Jupiter's Satellites

A primary human psychological urge is to expand. In order to accomplish expansion in a harmonious way—that is, according to Universal Law—an individual must approach the world in a balanced way. In examining the placement of the Jupiter's moons and their aspects, we gain a clearer understanding of how this planet works in our lives.

Keywords for Elements/Alchemical Actions

Io	*Europa*	*Ganymede*	*Callisto*
Fire	Air	Water	Earth
Potential	Binding	Form	Substance
Purpose	Intellect	Emotion	Body
Discriminating Awareness	Accomplishing Action	Mirror-like Wisdom	Equanimity
Passion	Envy	Anger	Pride
Inspiration	Competition	Emotion	Patience
Intense Intuition	Detached Thinking	Defined Feeling	Solid Perception
Calcinatio	*Sublimatio*	*Solutio*	*Coagulatio*
Power to Transform	Power to communicate	Power to Sustain	Cohesive Power
Spirit	Mind	Soul	Body

Examples

Elisabeth Kubler-Ross

Elisabeth Kubler-Ross' life work addressed death and dying, two of humanity's most difficult transitions. She endeavored to make the final phase of existence as powerful and rich as possible. Jupiter in Aquarius deals with expansive intellect that attains objectivity through the process of thinking. Further, Kubler-Ross synthesizes ideas from different sources and even different realms of mind, making her insights precise but not overly emotional.

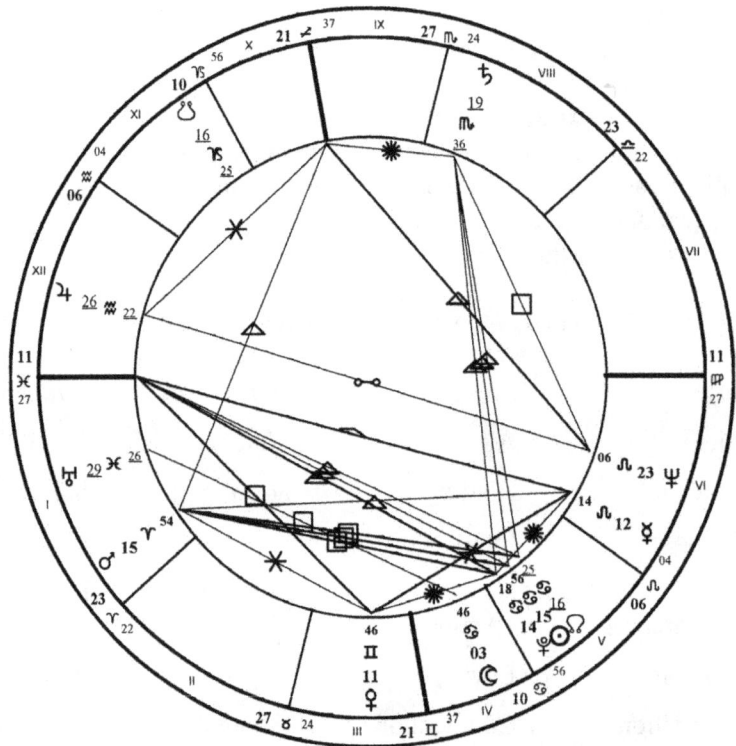

Elisabeth Kubler-Ross Geocentric Birth Chart
Birth data: July 8, 1926, 10:45 pm CET, Zurich, Switzerland

Using the Jupiter-centered chart, we explore the moons to gather information about Jupiter's function from the Jupiter-centered perspective.

Io is in Aries. The fiery alchemical energy is very personalized for Kubler-Ross and will express itself strongly through her persona. Dynamic expression of her ideas serves her well in presenting new pathways for relating to the end of life. Creative principles are provided in her work for all people concerned at the time of a death.

Europa is in Scorpio. Europa governs the air function in alchemy, the ability to objectify life's experiences. Here we see that Kubler-Ross had unusual skill in finding an objective perspective on death, enriching the transformative moment for all the people involved, and talking or writing about death in terms others can understand. She seemed to have the ability to look beyond the boundary of her own death in working with others.

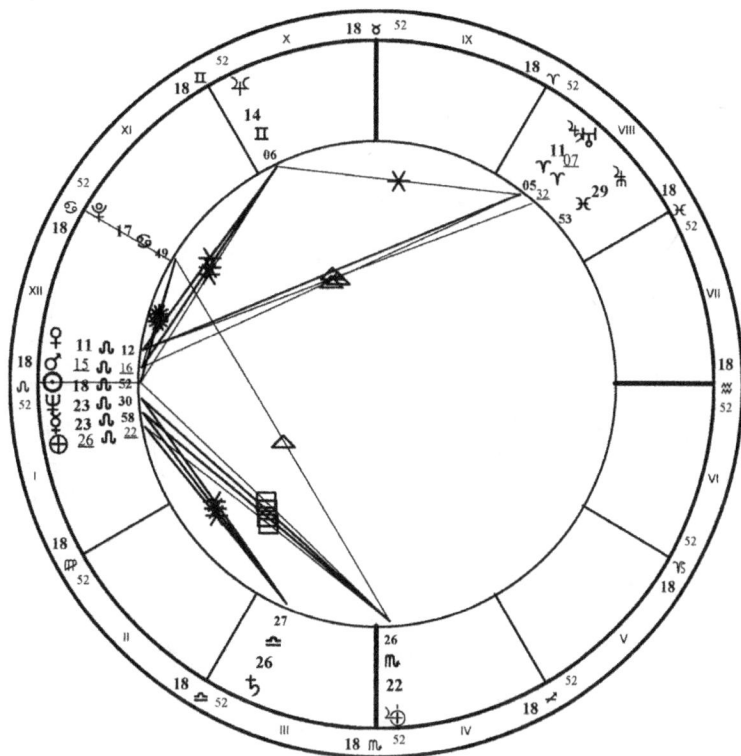

Elisabeth Kubler-Ross Jupiter-Centered Birth Chart

Ganymede expresses the water alchemical process of dissolving the existing structure within the psyche. Placed in Gemini, the sign of multiplicity and diversity, Ganymede allows Kubler-Ross' thinking function to dissolve old patterns of thought around death and dying. Because she is a feeling type (Sun in a water sign), the placement of Ganymede serves to reinforce a direction that is clearly shown in the natal chart, focusing on a specific field of study.

Callisto represents the earth element in the chart. Alchemical earth serves to re-solidify the material involved in a process so we can assess the changes that have occurred, the relative purity achieved, and what further process is needed. Callisto is in Pisces, signifying that the feelings involved in working with death and dying work into meaningful outcomes.

Most people do not have the conscious ability to work with all four elements or alchemical processes on demand. Usually at least one is less conscious, arising on its own. To the extent that Kubler-Ross is a feeling type, her thinking function may have worked on its own schedule, so to speak, potentially causing some irregularity in presentation or in the writing process, for example. Yet the placement of Ganymede in air and Europa in water will tend to ameliorate this kind

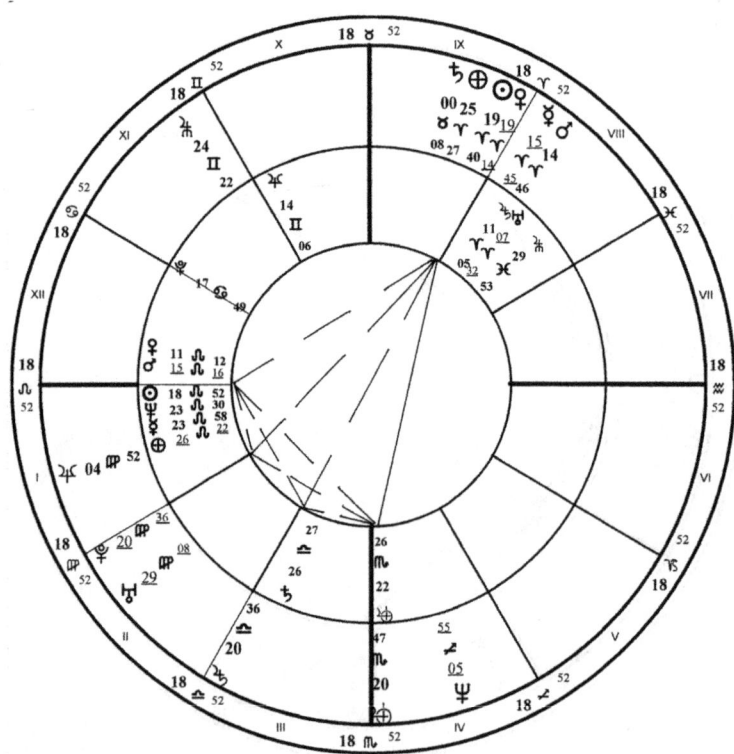

Elizabeth Kubler Ross (inner) and Publication of On **Death and Dying** *(outer)*

of problem, making Kubler-Ross' work more balanced. The spaciousness of her intellect provides another kind of balance as well.

In many ways we can view the publication of Kubler-Ross' *On Death and Dying* as the culmination of Kubler-Ross' life work. Although the book was published long before her own death, she changed the face of aging and death as it is treated around the world. She developed a model of the stages of the grief process that remains a cornerstone of psychotherapy, providing a reflective process for dealing with any type of loss.

On November 21, 1969, *Life* magazine published an article that thrust Kubler-Ross' work into the public spotlight for the first time. Note that the moons of Jupiter are moving very fast. Io moves about 8 degrees per hour, Europa about 4 degrees, Ganymede about 2 degrees, and Callisto just under 1 degree per hour. During the day of publication, Io moved into the midpoint of a Yod of birth Europa, transiting Pluto, and transiting Sun/Venus in the ninth house of publishing. These planets all aspected her birth Sun and Neptune. In addition, Ganymede was conjunct Earth in the birth chart early in the morning, transiting Earth was trine natal Earth, and

Callisto was sextile natal Earth. Transiting Mars and Mercury were sextile natal Ganymede and trine natal Mars. Transiting Callisto was sextile natal Mercury and Neptune. All these aspects occurred between 6:00 am and noon, focusing primarily around 10:00 am.

Considering the transiting aspects of Jupiter's moons, we have the following interpretations for the aspects to the birth chart:

Io Semi-sextile Europa: Passionate intellect stimulates phenomenal good fortune in your business dealings. Brainstorm with a close friend or associate for stunning results.

Callisto Sextile Neptune: When you go on an exploratory adventure, you bring two huge assets: your natural psychic abilities and your creative talents. And you DO love an adventure! You want to accept every chance to travel, work, or play.

Callisto Sextile Mercury: Always alert for the best way to resolve differences, you try to uncover additional information to bring opposing sides together.

Callisto Quincunx Europa: You take pride in intellectual capacity that provides a framework for mystical endeavors. Although you have deep interest in spiritual and mystical studies, your accomplishments in the world are often based on more mundane factors. Adapt your thinking to the ordinary needs of each situation.

Europa Conjunct Europa in Scorpio: You bring the power of communication into each moment by showing optimism in your work and other activities. Reveal your sources of information and share personal experience openly. Then you can build success that will be remembered for a long time.

The surest way for you to accomplish goals is through occasionally ruthless pursuit of desires, balanced by enjoyment of success. Avoid the temptation to get ahead by undercutting others. Logic tells you that if they are right behind you, they can more easily stab you in the back of they feel slighted.

You can utilize competitive spirit to balance your tendency to overrate your personal skills and abilities. Test your abilities in public to see how you measure up to others.

Europa in Scorpio captures the essence of Kubler-Ross' work and her intention for *On Death and Dying*. She states throughout her work that hope is essential to the dying process and should never be taken away. Europa in Scorpio in her birth Mars-centered chart is reinforced by the transit, cementing the concepts of optimism, clear communication, and enjoyment of success. She must have been more than merely gratified to see her work come to completion in this way.

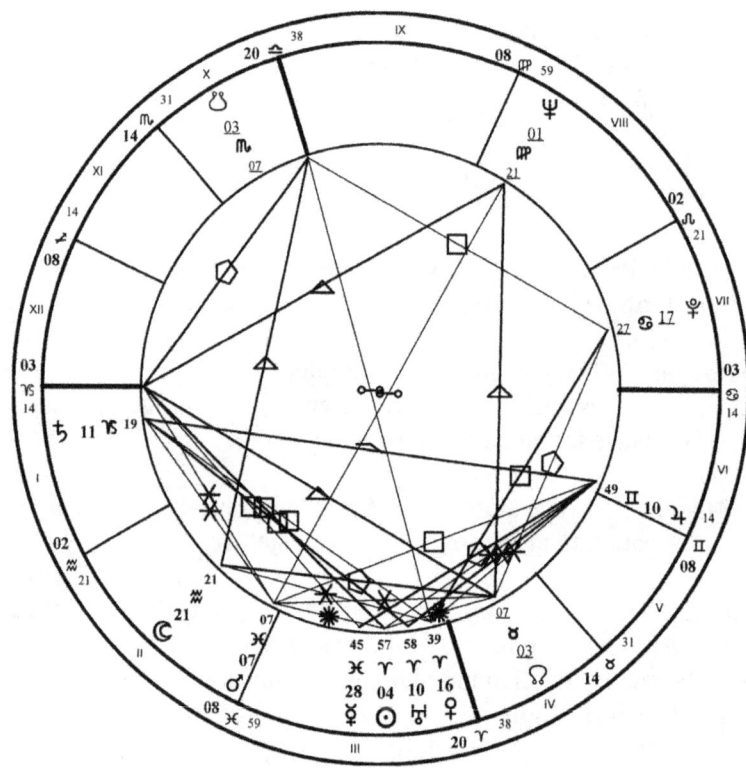

Sandra Day O'Connor Geocentric Birth Chart
Birth data: March 26, 1930, 1:10 am MST, El Paso, Texas

Transiting Ganymede Quincunx Birth Uranus (at 11:00 am that day): Along with your intuitive insight, you bring your emotions to the table when you participate in social and career activities. If you want to become an intimate member of the group, you may have to adjust the flame that fires your emotions.

It's easy to see that this day was filled with emotions ranging from the pride of accomplishment to the joy of exploration and ruthless pursuit of Kubler-Ross's aims. Intuition played a dramatic role in the development and culmination of her work.

Sandra Day O'Connor

Sandra Day O'Connor was the first woman member of the Supreme Court of the United States. Nominated to the court by Ronald Reagan in 1981, she took her seat on September 25 after her September 21 confirmation. In her birth chart, O'Connor has Jupiter in the sixth house in Gemini. Aspects of Jupiter include a quintile to Mercury (4' orb), a sextile to Uranus (8' orb), and a quincunx to Saturn (30' orb). Clearly Jupiter holds a significant position in her chart.

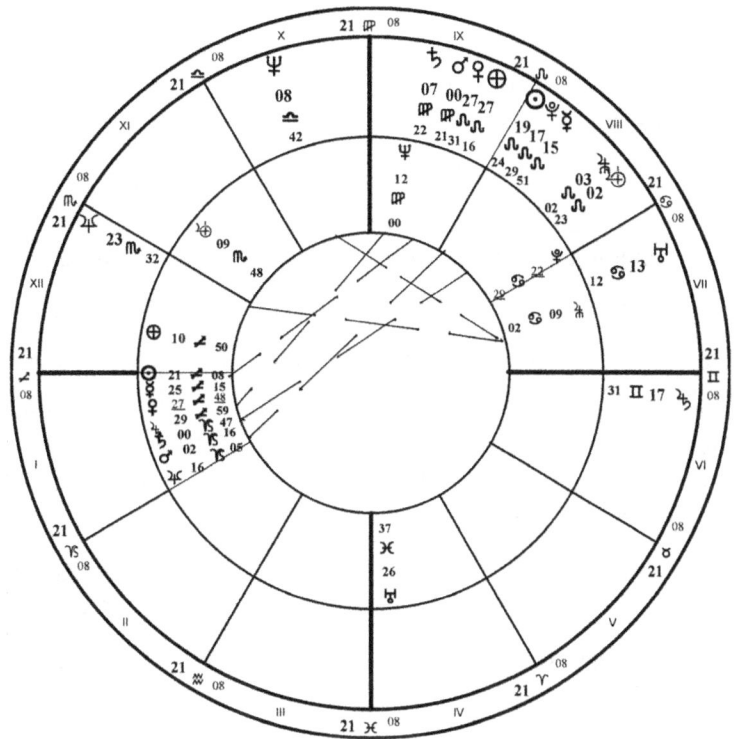

Comparison of Sandra Day O'Connor to her Successor, Samuel Alito
Jupiter-Centered Chart
Samuel Alito birth data: April 1, 1950, 12:56 am, Trenton, New Jersey

On December 12, 2000, O'Connor ruled with four other justices on Bush v. Gore to stop the Florida recount in the presidential election. In the Jupiter-centered chart for O'Connor for that day, Mars was trine her birth Pluto, indicating the power embodied in the ruling; the decision was split five to four along party lines.

O'Connor submitted her resignation from the court on July 1, 2005. Roberts was supposed to succeed her, but Roberts replaced Justice Rehnquist when he died before O'Connor's resignation became final. She was then succeeded by Justice Samuel Alito.

Examining the synastry between O'Connor and Alito in the Jupiter-centered biwheel chart, we find a number of very close aspects. Alito's Ganymede is sesquisquare O'Connor's Callisto, his Europa is quincunx her Mars, his Neptune is square her Callisto, and his Mars is trine her Saturn. Their Venuses are trine each other. The closest aspect is his Mercury sesquisquare her Saturn, reflecting her reason for resigning (she wanted to care for her dying husband).

Summary

Jupiter-centered charts offer significant information about function on all levels of experience. Careful study of these charts adds richness to our understanding of Jupiter's role in the natal chart, and to Jupiter's qualities in general.

The moons suggest which of Jupiter's facets will be more prominent in the individual chart; the elements and signs of the moons indicate the expression of Jupiter's energy that is most available to the individual.

1. Consider the Earth's sign/element. How does this reflect Jupiter's view of your wisdom and the role of love in your life?

2. Consider the sign and element of each of Jupiter's moons. List the possibilities for Jupiterian action that are naturally available to you. Think of a time when you used these choices, or when these choices might have worked better than what you actually did then.

3. Considering Jupiter in your geocentric chart, which of the moons in the Jupiter-centered chart reflects your typical Jupiter energy use?

4. How can you become more thoughtful in terms of expansion, based on the positions of Jupiter's moons?

5. Which element is stronger in the Jupiter-centered chart? How does that match up with the element Jupiter occupies in your geocentric chart?

6. What do the four moons and their sign/element positions suggest about how yu process information about the events and situations in your life?

7. Do you find a different chart pattern or different aspect patterns in the Jupiter-centered chart? With more objects in the chart, the patterns can seem more complex. Sometimes the moons and planets bunch together closely. Which is true in your chart?

8. Consider each Jupiterian moon as a skill or talent that you naturally have for using Jupiter's energy in your life. Make it a point to try out each "style" or process.

The Jupiter-centered chart reminds you of at least four distinct ways you can move the our your world effectively. In the next chapter we turn to Saturn and discover five ways to assess and use structure to enhance our knowledge and experience.

Chapter Six

Saturn-Centered Charts

In Babylon, Saturn was called Ninib and was an agricultural deity. Called Cronus by the Greeks, Saturn and Rhea (Saturn's consort) were agricultural helpers at the dawn of the Ages of the Gods. Saturn became ruler of the Universe after he eliminated his father Ouranus. It was prophesied that one day Saturn would also lose power when one of his children would depose him. He therefore endeavord to slay all of his children.

Through deceit Rhea concealed Jupiter from Saturn to save his life. Eventually Jupiter vanquished Saturn, who fled to Rome to become a favorite god there. Beginning on December 17 of each year, the Romans held a festival known as the Saturnalia. During this period of goodwill, freedom for a day was given to slaves, who then had first place at the family table and were served by their masters.

Given that Saturn is the slowest of the visible planets, he was thought to be the keeper of time, and this is why time is often depicted as an elderly man with a long gray beard. Although not the same as Chronos, the god of time or the ages, Saturn (Cronus) is associated with time to this day.

El, a Semitic god, has similar traits to Cronus. An old man wearing bull's horns, El was the father of many deities, including Dawn and Dusk. A creator god, this Canaanite deity caused rivers to flow. His consorts were Astarte and Anat. El fathered a king of Sidon and Mot, god of

Death. Anu, a Hittite god sometimes called the King of Heaven, also had many traits in common with Saturn.

Saturn in Astrology

Saturn defines the structure of our lives in terms of its cycle. Another focus, at least in the recent past, has been all the uncomfortable, "malefic" facets of Saturn's energy. Both these approaches seem to miss the mark in terms of the alchemical nature of this planet. Saturn deals with the structure of the physical, mental, and spiritual realms, and our perception of Saturn's influence can be negative. However, these are not all, nor even the principal expressions of Saturn.

Saturn stands as the boundary between the inner planets and the more recently discovered outer planets. This balance point, for the present state of humanity, represents intelligence. The human state involves the expression of intelligence through several vehicles, the primary one in the western world being logic. Science in the West has focused on mathematical logic, and anything that does not fit into this pattern has been set aside as being unreasonable or irrational. Intuition and instinct have become less acceptable. Many astrologers are in tune with instinct and intuition and are, along with psychotherapists and mystics, prime movers in the movement to reintegrate these necessary functions into consciousness.

Alice Bailey, in *Esoteric Astrology*, addresses the Saturn dilemma.[21] Western astrologers have come to see Saturn as crystallizing and fixed, therefore limiting and restricting, whereas Bailey speaks of Saturn as a planet that "breaks up existing conditions by the force of its energy impact." Here we discover that structure is the basis for Saturn's expression.

Structure exists in the universe and is neither good nor bad. Structure seems bad when its expression becomes rigid and inflexible, as with arthritis or other symptoms of the aging process. Mental rigidity occurs when the ego complex cannot expand or flex to accommodate new information. Emotional rigidity defends against experiences that cannot be absorbed into consciousness. Spiritual rigidity is experienced when religious or philosophical forms become more important than the principles behind them.

We also experience structure as necessary and good. Necessary structures include bones and teeth, buildings, language, and thought patterns that allow us to communicate and think logically, and feelings that provide a medium for evaluation and judgment. To the extent that these structures are consistent with our wants and needs, they are experienced as being good. They only seem bad when they don't serve our needs.

While we often experience the crisis of Saturn as too much rigidity and too little structure, the crisis actually involves a point of balance. Saturn is strong in Libra. The most exalted expression of Saturn is to make a sound decision based on a clear understanding of the available choices. These choices serve us best when they include a balance of logical options with emotional choices frequently associated with unstructured, irrational behavior and thought.

In many depictions of Saturn in the Tarot and astrology, Saturn is a Devil to be "cast out" or exorcised. In actual practice we cannot eliminate Saturn because without his energy we will find ourselves incomplete. Retaining a conscious experience of Saturn's energy promotes balance within the psyche. Thus we need to acknowledge Saturnian energy for what it is and give it its rightful place.

In astrology Saturn is traditionally depicted as a discordant, malefic, cold, dry, magnetic, negative planet. Many astrologers mention Saturn transits as though they are a plague and even joke about what life would be like without this planet in their charts.

The physical expression of Saturn energy is fundamental to our existence; it rules bones, teeth and skin. Saturn provides us with form from the inside out physically. Saturn also reflects the logical, organized form our thoughts take, shaped as they are around language. Saturn reflects the energy of concentration, and this trait is one way we learn.

Emotionally Saturn is often experienced as fear. Wherever we feel frustration or limitation, we can find Saturn at work. Saturn is the lord of social, personal, and emotional boundaries, and often our experiences of structure determine our optimism or pessimism.

Positive expressions of Saturn seem to take time to grow and become comfortable, rather like old shoes. Saturn's energy is the basic stuff of security; in many ways, comfort ultimately depends on Saturnian structure. We don't find comfort in constantly changing situations.

Saturn's goal is perfection, but what is perfection, after all? Alchemical processes remove impurities so we can approach perfection. When our lives become limited and rigid, we say it is Saturn. When we find our studies oppressive, we experience Saturn's energy. When we are blocked, we blame Saturn. Actually it is our own one-sided nature that creates the experience of limitation and oppression, not Saturn's supposed tendency toward blockage.

Saturn has been named the Lord of Karma, but I believe he is rightly the Lord of Dharma. Karma is both the present effect on your life of past actions on the future predictable effect of present actions. Dharma, on the other hand, involves duty to yourself and your own inner nature, not to someone else's idea of who you are.

Saturn represents durability. After any physical or mental change process, we respect the strength of the human form. Sometimes we wonder if we can make it through. It is no accident that alchemical books are filled with prayers between descriptions of the process! To the extent that we are open and curious about change, we can adapt our lives to present needs without remaining stuck to old forms.

Saturn Key Words

Physical	*Emotional*	*Mental*	*Spiritual*
Skin	Boredom	Ambition	Karma
Bones	Depression	Common sense	Discipleship
Disease	Persistence	Foresight	Shiva
Endurance	Worry	Hindsight	spiritual crisis
Teeth	Seriousness	Concentration	Dharma

Isn't it interesting that Saturn is so ill thought of by many astrologers, considering that it is made up of our projections? Psychologically, we project that which is unconscious and we place in the unconscious the things that make us uncomfortable. Therefore, we can see that projections are made up of whatever has been uncomfortable in consciousness. This is consistent with what we think of Saturn—limitation, dark forces, stern parents, heaviness.

Saturn's Environment and Satellites

Saturn and his moons go beyond limitation to demonstrate the creative structure in our lives. The rings and satellites of Saturn form one of the most beautiful astronomical sights. The most distant of the planets visible to the ancients, Saturn became more and more interesting as telescopes were developed.

Saturn and Earth share nearly identical inclination of their axes, with Earth's at 23.5 degrees and Saturn's at 26.73. At the equator Saturn is 119,300 kilometers in diameter. It has 744 times the volume of Earth, yet has a mass only 95.17 times that of Earth. This relative lightness results from Saturn's composition of gases and a very small metallic core. Winds up to 500 meters per second combine with a surface temperature of -180 degrees Celsius. The rotational period (day) is 10 hours 39.4 minutes; the sidereal period (year) is 29.46 earth years.

By 1980, seventeen moons of Saturn had been identified. The satellites range from immense size to fewer than ten miles in diameter. There are systems of moons that share the same orbits, maintaining equal angular distance from each other. (Tethys is one of these; the astrological impact of the two co-orbitals, Calypso and Telesto, will be discussed later.) Others have gravitational effects on the co-orbiting planet; these "boost" each other into higher orbits from time to time, with the companion falling closer to the planet. (Dione shares such an orbit with Helene).

There are no true eclipses on Saturn. Iapetus is the only moon whose angle of inclination provides a view of Saturn's rings; the other moons' orbits lie very close to the plane of the rings.

Delineation of Saturn-Centered Charts

The Intrepid astrology program tracks the five largest moons of Saturn. These five provide interesting mythologies that parallel their astrological significance.

Iapetus

Iapetus (satellite astronomical number VIII) has a period of 79.331 days. One hemisphere is far lighter than the other. Suggestions for the cause of this difference include deposits from Phoebe, another small Saturnian moon, or upwellings from the interior of the moon. Like other major satellites, Iapetus has a synchronous orbit—the same side of the moon always faces the planet. Features on the surface—craters and other regions—take their names from the *Chanson de Roland*.

In Greek myth, Iapetus was the father of Atlas and Prometheus, as well as Menoetius and Epimetheus. Both Atlas and Prometheus became known for their perseverance under great stress; Atlas bore the weight of the world on his shoulders, while Prometheus suffered endless punishment for stealing fire from the gods. This slow-moving moon is thus rightly connected with the earth element.

Iapetus's position in the Cronocentric chart will indicate the nature of a satisfactory outcome or product. Iapetus generally seeks measurable outcomes and will state even illusive spiritual or emotional outcomes in concrete terms, as in "I feel strong as an oak," rather than merely "I feel better."

Titan

Larger than the planet Mercury, Titan (astronomical number VI) is the largest of Saturn's moons. With a period of 15.495 days, Titan is enveloped in an orange cloud of atmosphere composed primarily of nitrogen and methane. All the ingredients necessary for life exist on this satellite, but the temperature of -186 degrees Celsius seems to forbid it. Some astronomers suggest that Titan has a solid methane surface, rivers of liquid methane, and an atmosphere of methane, all of which are indeed unappealing from a human perspective.

In mythology, the Titans were giants who populated the early world. They were a divine race responsible for the invention of arts and magic, and the ancestors of men. Our modern language retains their name in words like titanic and titanium. This moon of Saturn (himself a Titan) is larger than all moons save Ganymede.

Representing the watery alchemical process of coming to form, Titan reflects desire. The watery process seems antithetical to form, as it dissolves the existing *materia* in order to purify it. Water does not have its own personality but flows to fill whatever container is provided. Yet water also dissolves the container over time. Calm water has a crystal clarity to it.

Saturn's rulership of time is seated in the flowing water element represented by the moon Titan. The sign occupied by Titan defines the quality of change within the structure of Saturn's role in the natal geocentric chart.

Rhea
Discovered by Cassini in 1672 and bearing the astronomical number V, Rhea has a period is 4.518 days. Larger than either Dione or Tethys, Rhea has a surface similar to the others. Like Iapetus, the trailing half of Rhea is rather dark. This satellite appears to be very ancient with rubble covering the cratered surface.

Rhea, daughter of Uranus and wife of Cronos (along with her sister Dione), had three daughters and three sons. Cronos swallowed the first five to prevent the fulfillment of a prophecy that one of his children would overthrow him. Rhea hid Zeus when he was born. Later Zeus did overthrow Cronus, freeing his siblings Hestia, Demeter, Hera, Hades and Poseidon.

Rhea, not surprisingly, signifies the quality of learning from experience. Personal ability to rise above experiences can be read in Rhea's sign placement:

- Fire: sincerity, creativity and the use of intuition.
- Earth: capacity to concretize objectivity in some way.
- Air: ability to communicate, often through writing.
- Water: capacity for emotional change along with relative objectivity.

Rhea goes beyond intellect and understanding to reach another level of understanding that is essential for individual objectivity.

Dione
Dione, numbered IV by astronomers, is close to the size of Tethys, but of denser material. One side of this moon is substantially brighter than the other.

Dione has a co-orbital satellite, Helene. Unlike the shared orbits of Tethys' companions, Helene and Dione interact gravitationally, causing mutual changes in orbit and speed. Thus the two effectively exchange orbits. The surface of Dione is marked by craters, chasms and lines that may have been formed, in part, by material breaking through the surface from within the moon. Dione's period is 2.737 days.

Dione is a feminine name corresponding to masculine Zeus. Wife of Cronus, sister to Rhea, and mother of several girls, most notably Aphrodite, Dione was connected with Zeus' cult at Dodona.

The creative aspect of fire associated with Dione cannot be over-emphasized as it indicates the inspiration for the first essential step in the overall process of coming to form. Dione's placement in a Saturn-centered chart defines individual creative inspiration, thereby suggesting the ultimate impact of one's inspiration on the final outcome of creative processes.

Just as Dione does not orbit Saturn alone, the fire element cannot produce concrete results by itself. When analyzing Dione's placement in a Saturn-centered chart, one must consider her potential in all the signs, as the co-orbital Helene may occupy any sign in this chart. Thus the creative fiery process for Saturn is a considered one, not indiscriminate in its operation. Dione suggests the underlying inspiration with which all structure is invested.

Tethys

Designated "III," Tethys is smaller than the four moons already discussed, Tethys' orbital period is 1.888 days. Two very small satellites, Telesto and Calypso, share Tethys' orbit, occupying the Lagrangian points 60 degrees ahead and 60 degrees behind. The surface of this moon has one huge crater, another that is quite large, and a unique trench that runs from the north pole of the moon through the equator, ending near the South Pole.

One of six female Titans in Greek mythology, Tethys was daughter of Uranus and Gaea, and spouse of Oceanus. She was the mother of 3000 sons, the rivers, and 3000 daughters, the water nymphs (oceanids). Together with Oceanus, Tethys raised Hera and educated her. Tethys' daughter Calypso personified the depths of waters.

Tethys has a nature outside the four elements, and this etheric energy suits Saturn's energy well. Tethys in her ocean is used to vast space. She indicates how Saturn provides space for the individual and how intelligence may take form. Calypso and Telesto confirm that no teacher works alone. Calypso encourages the student to look deeply into a subject; Telesto offers the opportunity to achieve breadth of subject interest. Tethys indicates how an individual develops wisdom; Calypso and Telesto indicate sympathetic energies available to the individual.

Examples

R. Buckminster Fuller

Fuller wrote: "You cannot get out of the Universe. You are always in Universe. . . . Nature coordinates in twelve alternatively equieconomical degrees of freedom—six positive and six negative. For this reason, twelve is the minimum number of spokes you must have in a wire wheel in order to make a comprehensive structural integrity of that tool."[22]

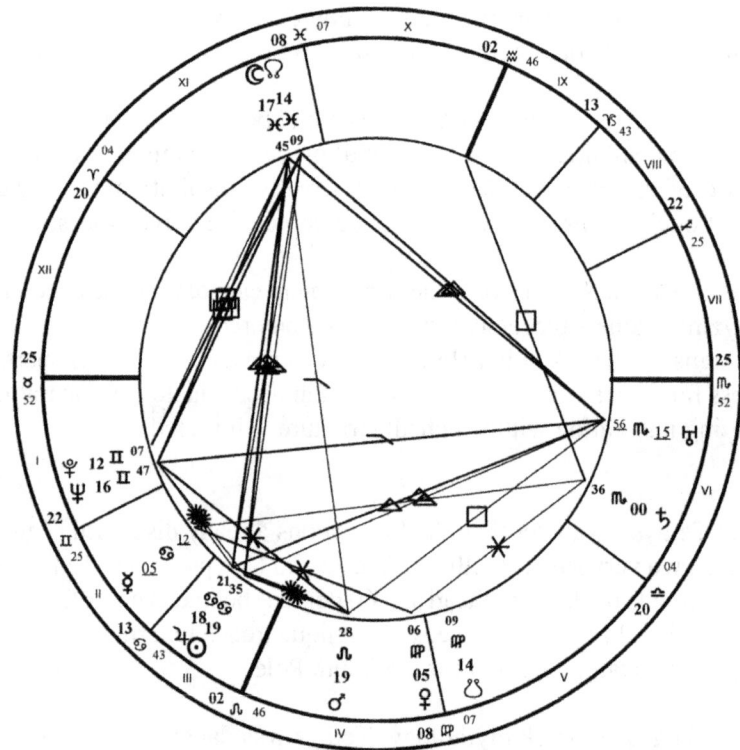

R. Buckminster Fuller Geocentric Birth Chart
Birth data: July 12, 1895, 12:45 am EST, Milton, Massachusetts

In the above quotation we find reflected the astrological chart and the structure of the zodiac. Fuller saw space as the absence of something rather than a thing in itself. He needed something tangible to define reality for himself and arrived at a conclusion strikingly astrological: twelve is a proper number to make the material world work. He even perceives that world in equal halves, positive and negative. Nowhere does Fuller intimate his understanding that this is a fundamental astrological principle, but rather he uses worlds like "omnirational" to describe his universe.

Fuller's Saturn-centered chart reveals the "how" of his work. In this chart the T-Square involves Tethys, Titan and Earth—a richly dynamic pattern. In Fuller's transpersonal world, therefore, Saturn's structured capabilities are fully active. With Tethys and the Earth in earth signs, one of the musts from Saturn's perspective is material expression. Fuller's talent manifested in his engineering work and fulfilled the Saturnian promise without the necessity of having Saturn directly involved in a strong geocentric pattern like his Grand Trine.

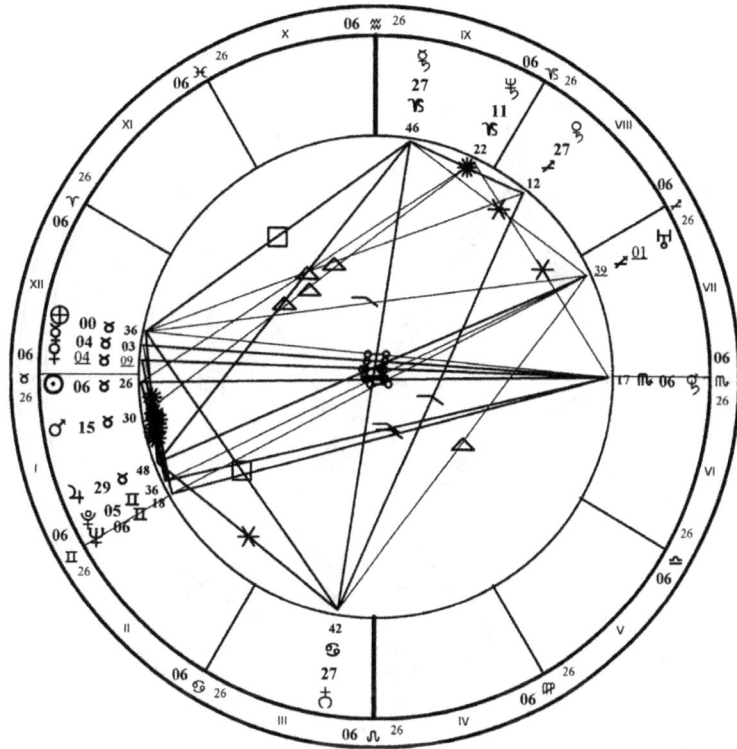

R. Buckminster Fuller: Saturn-Centered Birth Chart

Saturn's moons outside the T-Square address unique qualities in Fuller's work. Fuller himself talks about individual limits of understanding because of physical over-specialization and the narrow electromagnetic spectrum range of our vision. For example, to stimulate the creative connection and make it more evident, he chose philosophical slants for his writing while allowing ideas to germinate at their own speed.

Rhea is trine the Earth, indicating Fuller's capacity to manifest his intellect productively. His humanness itself allowed him to bridge the gap between thought and expression; the same quality made him aware of the limitations of his audience as well.

The extremely close aspect between Tethys and Titan emphasizes the degree of Fuller's awareness of structure in his environment. This opposition is a core aspect in a mystic rectangle of Tethys, Uranus, Titan and Jupiter, showing the essence of intuition and expansive thinking throughout his life and work.

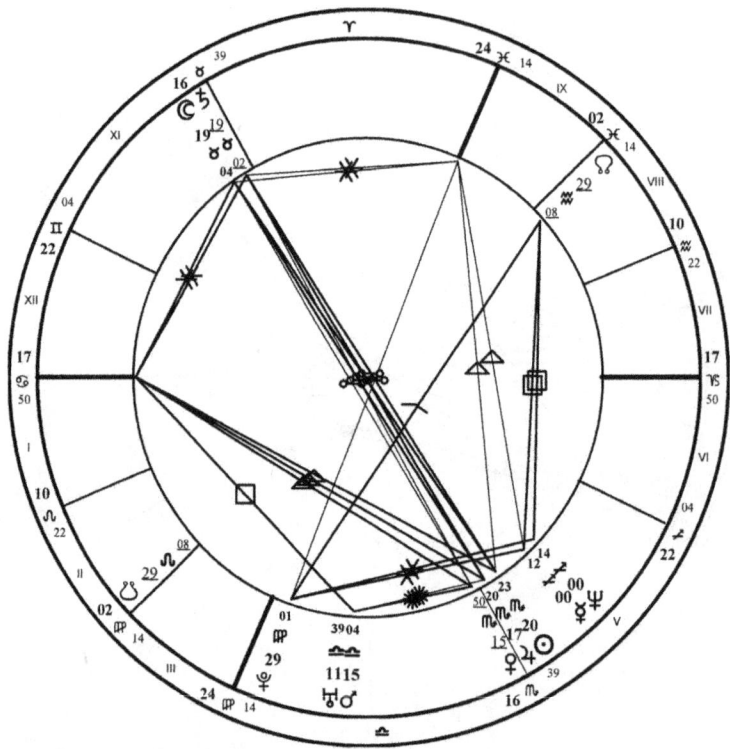

Tonya Harding Geocentric Birth Chart
Birth data: November 12, 1970, 8:22 pm PST, Portland, Oregon

Tonya Harding

Tonya Harding rose to the top of the figure skating world on the basis of her talent. Notably, she was the first female skater to land a triple axel jump in competition. She fell from her pedestal because of associations with people who may have acted out her darkest desires. In her birth chart, Saturn is exactly conjunct the Moon, both opposing the Sun, Jupiter, and Venus. Titan, Saturn's planetary companion, opposes Uranus and Mars. The oppositions define Ms. Harding's "either/or" approach to life. The Saturn-centered chart often indicates potential relief from one's dilemmas. However, in Ms. Harding's chart, Titan is tightly square Dione and opposition Pluto, reinforcing the harsh quality of the Saturn and Titan aspects in the tropical chart. The following interpretations indicate how she might have managed Saturn's energy more successfully.

Titan in Aries: You have a wealth of intuitive data sources. When you pay attention to your intuition, you no longer need to press forward in an ego-centered way. Instead you sell your ideas based on their practical values.

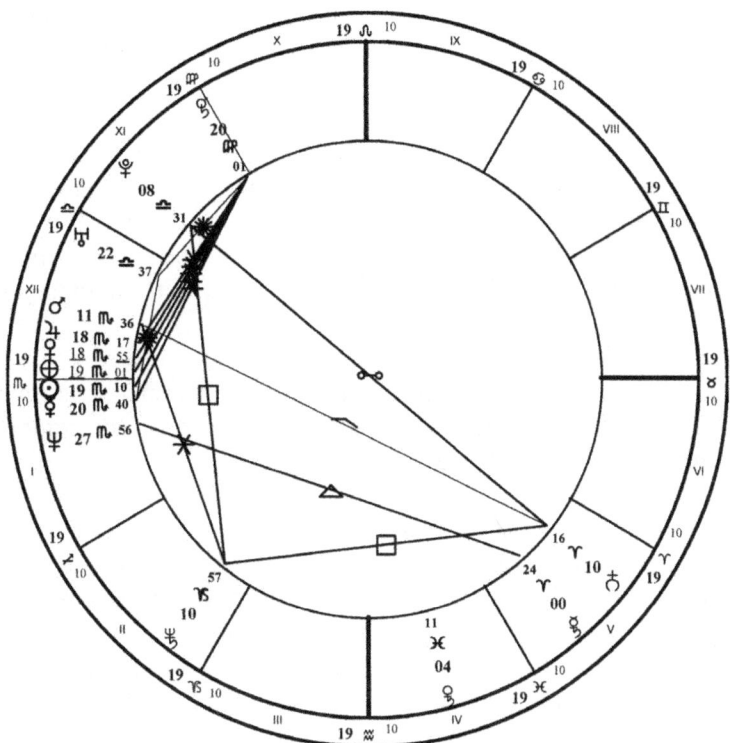

Tonya Harding Saturn-Centered Birth Chart

Tethys in Aries: Focusing on the collective needs may not come naturally to you in the beginning. Your natural style is more independent. The same skills that make you independent make you a strong leader. At times you feel inhibited in crowds. Your enthusiasm for your task carries you through these moments.

Rhea in Pisces: Logic provides a solid tool as you undertake the task of understanding and integrating your values. Study more than one religious or philosophical system. Conservative thinking may give way to new insights.

Dione in Capricorn: Loyalty is a quality you value greatly in your associates. However, it is nearly meaningless if you don't express it yourself. Loyalty is only earned when you exert it yourself. Selfishness, a character trait you believe is part of the human equation, only becomes a problem for you if you let it rule your life.

Iaptetus in Virgo: Industriousness could be your most natural capacity. You actually enjoy being busy and having new projects to look forward to.

Titan Opposition Pluto: You have opportunities in the scope of daily life to more effectively use your power and will. Awareness, for you, increases as you engage in daily life. Then your psychic talents come into play naturally.

Titan Quincunx Mars: Do you find yourself fiddling with your laces when you should be stepping onstage? You benefit from careful, meditative preparation before any important activity.

Titan Square Dione: Your accomplishments on the material plane result from merging intuition with your emotional sensitivity. Feelings sometimes prevent you from finding unique solutions to odd questions.

Tethys Trine Neptune: Devotion flows into you through an emotional or psychic channel. Then you must evaluate the emotional content using analytical methods. You naturally merge devotion and thought into creative activities of all kinds.

Dione Sextile Mars: Creative energy flows when you plan from the outset to bring all aspects of a problem into the solution. Be fearless in accepting chances given to you to share activities with associates. The group energy enhances your efforts to bring separate elements together.

Dione Square Pluto: Powerful inspiration comes to you when you remain open to the unusual energies all around you. Your life revolves around challenges that draw upon your psychic senses as well as your intellect.

Iapetus Sextile Jupiter, Venus, Earth, Sun, Mercury: Because you can often visualize the outcome of your actions, you challenge yourself to work for the highest principles of wisdom and love. Use every opening presented by others to test your ability to construct and execute your plans.

You begin relationships with the expectation that they will last. You would not waste your time on anything less. You often see possibilities for the future that your friends overlook.

Your mission to communicate in the world depends on how you begin both studies and their application. Use every chance you get to talk with other people who share your interests in making change happen.

Your mission, should you choose to accept it, is to manifest new directions for the people around you. You can become a powerful manager because you follow the thread of each opportunity at least a little way to see where it may lead. Then you make your choices.

Summary

By focusing on structure as a principal quality of Saturn, we have formulated an alchemical model that shows solidity is only one aspect of apparent reality. Fuller and Harding both lived their lives as though the solid physical world really matters. The Moons of Saturn reveal five natural choices each of us has for dealing with what seems all too rigid at times.

1. What does the concentration of planets around the Sun suggest to you in terms of how Saturn might see your understanding of structure?

2. What elements/signs have strength, based on the position of the inner planets?

3. What elements/signs have strength, based on the positions of Saturn's moons?

4. How do the above two considerations compare or contrast to the position of Saturn in your geocentric chart?

5. Is there a Saturnian moon in the group around the Sun? What does this suggest to you?

6. Consider the chart pattern and any prominent aspect patterns. If the chart seems overly complex, you may want to narrow the orbs of aspects. However, remember that the moons move very quickly, so allowing wider orbs allows for a small error in birth time.

7. Which of Saturn's moons occupies a position that seems particularly significant or impressive to you? Is there one (or more) that suggest new ways of handling structure in your life?

The next chapter takes on Uranus, the personification of Change.

Chapter Seven

Uranus-Centered Astrology

Uranus represents change in all its forms, whether it be through a revolutionary shift of energies in the material world or a comparable rhythm within the psyche. We often perceive change as destructive, eccentric, and impractical because we don't want to experience change at all—we want things to remain the same because that provides the illusion of security within a rigid Saturnian structure.

Greek mythology pairs Uranus with Gaea as rulers of heaven and earth. Hindu myth links Mitra and Varuna as gods who "did not institute, but who maintain universal order."[23] Etymologically, the names Uranus and Varuna are connected and the relationship between these two expressions of the myth can help us understand the more positive, beneficial side of Uranian energy.

The names Varuna (Hindu) and Uranus (Greek) both derive from the noun for heaven. The term also means master of the bond, a role that Varuna seems to have developed more fully. Metaphorically this bond can be seen as the principle of Uranian energy—forward movement of all things that prevents us from stepping out of the current of the universe.

Varuna is the closest Asiatic parallel to Uranus. God of the sky and also of providence, he is the source of the great rivers of the world, has the quality of eternal vigilance, and exacts payment for any evil which he discovers. As he is able to observe everything, this god directs the physical and moral world, rewarding and punishing, but also taking into account one's intention and insight into any wrong actions that may have occurred. This potential for mercy reveals the duality which we can expect from Uranus.

In Hindu cosmology Varuna was associated with an abstract principle called *rita*. This principle indwells in everything in the universe that has the quality of regularity or order in it. Thus we can presume that Varuna indwells in everything to the extent that it is part of a larger pattern; it is this pattern of things that contains the positive expression of Uranian energy for us.

A story about Varuna shows both sides of his character. A human king, Harischandra, had many wives, but no male heir. He prayed to Varuna, promising that the life of any male child would be offered to the god; Varuna accepted. However, when the male heir was born, the king used deceitful means to escape the promise, suggesting to Varuna that the child should first teethe, and later that there were imperfections that could be resolved with time. The son grew to manhood.

The king finally told the son of his promise to Varuna and the son fled. The king began to suffer for his unfaithful actions and the son attempted to resolve the situation by offering sacrifices to Varuna. The son of a Brahman was offered as a more suitable sacrifice because of his religiousness. In the end Varuna released both sons, but only because of their prayers; thereafter the god ceased to require human sacrifices.

Within the decision of Varuna we find the egoless quality of Uranus. The god demonstrated mercy because of the prayers of the two sons, not because of the Mars-like desire of the king. Whatever we experience from Uranus, it is found in an egoless expression of the forward movement of all things; any ego in the experience is our own projection.

Uranus ruled the heavens in consort with Gaea, who ruled the Earth. Between them they produced the Titans and numerous other offspring, among them Saturn. Saturn eventually dethroned Uranus, mutilating him in the process; we see this mutilation as a metaphor for what can occur when we attempt to resist change.

Going a step further with the myth, we can discover more metaphorical meaning. Saturn threw his father's genitals into the sea, producing a dual reaction. Some of the black drops of blood that fell produced the Furies, while the rest produced a white foam out of which Aphrodite emerged. Here we see that the dark underground of the psyche, represented by the Furies, came from the same source as the beautiful, feminine Aphrodite, an expression of the *anima*.

Isabel Hickey, in *Astrology, a Cosmic Science,* states that Uranus represents an energy about which you can do nothing. What things are there about which we can do nothing? For one thing, we cannot stop the forward motion of time. The world continues to move through time and we can only seem to delay that movement or postpone its results.

On the personal level Uranus governs bodily rhythms as well as the nervous system as it connects with other tissues. Time wears down the physical machinery of the body and the best we can do is to keep the body in an excellent state of repair. Beyond the physical improvement we experience through proper nutrition and care of the system, we can pursue psychological paths of self-discovery, mental growth through study, and spiritual development as well. Joining in the forward motion of the universe, we find that we cannot turn back—we must go forward or break.

Uranus is an agent for personal change on every level. Through breaking up old forms, new ones can emerge. The Self remains intact even through radical changes in our mental, physical, emotional, and spiritual forms.

Uranus has no definite form. It is not a distinctly masculine or feminine energy, although at times we experience one or the other. Uranus merges with other planetary energies, adjusting to whatever function is present at the moment, yet guiding the forward moving process all the while.

On the physical level we can make dramatic changes in our lives through exercise and nutrition. At first these changes may meet resistance from habits we have formed long ago. Psychotherapy utilizes Uranian energy to remove resistance to our inner psychic messages, both from the Self and from the Spirit. Of course no processes operate in a vacuum: the psyche will respond to healthy nutrition, just as the physical body will respond to a lightening of the emotional load.

Uranus not only rules change; it rules astrology itself. Uranus shows to us the profound potential of astrology to help people to be happier and more. As astrologers we utilize a body of knowledge older than any of the modern sciences. Astrology calls upon us to reach beyond the safety and security of our mundane existence to grasp insights concerning the very nature of the universe and its power. As Mercury reflects intellect and communication, Uranus reflects expanded intellect that draws upon intuition.

Uranus Key Words

Physical	*Emotional*	*Mental*	*Spiritual*
Rhythm	Independence	Intuition	Alchemy
Pituitary	Enthusiasm	Comprehension	Psychism
Irregularity	Zeal	Curiosity	Transformation
Nerves	Abruptness	Ingenuity	Transmutation
	Eccentricity		
	Psychology		

Equilibrium

It is my opinion that the most significant quality of Uranus is equilibrium. Throughout human history the gods associated with Uranus have been responsible for restoring equilibrium to the world and their actions take on new significance when viewed from this perspective.

The key principal for the delineation of Jupiter was function; that for Saturn was form. In researching the moons of Uranus, the principle of equilibrium is a central focus. Time after time, clients have complained about the loss of equilibrium during Uranus transits, yet Uranus' role is actually to reestablish balance.

Moons of Uranus

The perspective from Uranus includes numerous moons and interesting rings. For the purpose of this book on planet-centered astrology, five moons are included, all discovered in the 18th to 20th centuries: Oberon and Titania, discovered by Herschel in 1787; Ariel and Umbriel, discovered by Lassell in 1851; and Miranda, discovered by Kuiper in 1948. The order in terms of speed from fastest to slowest is Miranda, Ariel, Umbriel, Titania, Oberon, and the moons have been related to the elements in this order: space, fire, air, water, and earth.

The orbit of Uranus is remarkably stable with an orbital eccentricity of only 0.047. Uranus' path around the Sun is more uniform than that of Mercury, Mars, Jupiter, Saturn, or Pluto. The greatest latitude Uranus achieves is approximately $0°49'$ (in 2007). This means that Uranus' orbital path is very close to the ecliptic. It compares with Mercury at $7°$, the Moon at $5°17'$, and Pluto at $16°$.

When people describe the intuitive life, they are relating a Uranian process. The five Uranian moons provide a picture of different styles of intuition. Presented in this light, the interplay of alchemical energies offers insight into different intuitive potentials. By focusing on intuition and the judgement that can come from understanding, Uranus's moons offer their equally unusual ways of perceiving the world.

Symbolism of the Moons

The names of Uranus' moons, unlike the other planets, are drawn from the plays of Shakespeare and a poem by Alexander pope entitled "The Rape of the Lock." There is some confusion about the moon Ariel and the source of its name: Ariel is one of the chief actors in Pope's poem; he is also a spirit in Shakespeare's drama *The Tempest*.

Miranda is the smallest, innermost of the five moons. Before the first scene of *The Tempest*, Miranda and her father have been shipwrecked on an island. Thus, from the first moment of the play, they need to bring their lives back into equilibrium. Prospero invites a spirit to change events and to reestablish balance. Miranda is the human vehicle, while Ariel manifests the necessary charms.

Astrologically, Miranda represents the Void—the space in which equilibrium perpetually resides. The sign shows the first characteristic to emerge in the process of establishing or reestablishing rhythm after any upset. Miranda, the only human character to give a name to Uranus' moons, embodies the feminine quality of space; she also signifies the quality of intelligence at the root of all action in the universe.

Ariel can move at light speed to exercise his creative genius. Not infallible, Ariel is both quick and light. Thus the second Uranian moon represents the fire process. When imbalance occurs, Ariel shows the best inspiration toward change. While some people use fire to heat up situations, others may use the heat of the moment to move toward balance and harmony.

Umbriel represents the air element. The character in Pope's poem is only briefly mentioned, but I feel it captures the sadness caused by a misuse of intellect. We sometimes consider and reconsider a position, unable to move. Umbriel is able to fix Belinda's thoughts on the loss of her hair, and she becomes stuck in her sad thoughts. Umbriel shows where intellect seeks balance. Overworking intellect results in neglect of other processes. The placement of Umbriel may very well indicate energy to be used less, not more.

The last two Uranian moons, Titania and Oberon, are named for characters in *A Midsummer Night's Dream*. In this play the king and queen of the Fairies are key players in three plots involving the workings of love. Titania has the watery role of feeling. In the play she dreams of love. In the end she lets go of her dream image and return to her appropriate role. We sometimes face similar choices, giving up dreams for the sake of more balanced reality. Titania in the Uranus-centered chart shows us how to understand feelings in the midst of outrageous imbalance in thoughts and perceptions.

Oberon, the fifth moon, is driven by perception. He uses all his power to obtain what he wants through the manipulation of perception on the part of mortal and fairy alike. In the process he upsets both the natural and supernatural planes; however, he has the power to set things right again and does so. Early in the play Oberon experiences a wave of unconscious desire. Toward the end he settles down again, unwinding seemingly misdirected energies, leaving relationships the way they should be.

Uranus has a powerful impact in matters of love. This makes perfect sense when we recognize the importance of ceremony and ritual (Uranus-ruled concepts) in our interpersonal relationships. Although the word ritual is found within the word spiritual, the terms come from separate roots, *ritus* (rite) in the former and *spiritus* (breath) in the latter. Without the container of ritual, we become insecure about dramatic changes. Uranus provides ritual boundaries within which change can safely occur; the moons of Uranus reveal our individual talents for creating balance in our lives. More, they show where imbalance is likely to occur, providing us with conscious ground for learning life's lessons.

Uranus' role in restoring equilibrium is not gentle. It can be cosmic in its effects. It can also evolve into an intuitive balancing mechanism. The closer we are to balance, the greater equanimity we demonstrate, even in the face of major life-changing events. The further we are from the flow of our lives, the greater difficulty we have with life's adjustments. Readers may recall situations in which they or their friends experienced enormously traumatic events, but were able to move through them. Other times little difficulties assume mammoth proportions. The difference seems to me to be this sense of equilibrium.

The activity of Uranus in the geocentric chart often cues us to the level and kind of stress that accompanies life changes. The Uranicentric chart reveals the mind of the planet. If Ouranus, Varuna, or Kwannon is a force in our lives, we benefit from understanding something of his or her viewpoint; the natal Uranicentric chart reveals the radically different perspective from this first of the trans-Saturnian planet-centered charts.

Uranus, the Ray Seven planet, expresses "the will to be and to know simultaneously on all planes of manifestation." Through ritual, Uranus, the Ray Seven planet, expresses both the will to be and the will to knowledge. This will eventually is capable of expression on all planes of manifestation. The ritual magic required to achieve this level of awareness infuses itself into every human activity, as though we are completely dependent on ritual for our continued inter-relationship with others and with our inner Self. Some situations have obvious ritual processes, while others conceal some of the steps or disguise them. Suffering occurs when ritual processes go wrong or become overly obscured. Even the most difficult processes more forward with less pain when ritual steps are overtly honored. See Appendix Three for Jose Arguelles' steps in the ritual process.

The will to be and to know can well be satisfied through skillful use of language. Language is a basic ritual behavior. You can study rituals designed by other people for group situations and for individual meditative processes, yet ultimately you define personal rituals for yourself.

Exoteric astrology gives Uranus the role of sudden and abrupt change on the physical level, intuition on the mental level, and rhythm on the biological level. As astrologers we have studied the movement of Uranus and experienced the sudden changes Uranus can signal.

Intuition is often experienced as a rather sudden understanding, yet upon examination we realize that we have been gaining understanding over some extended period of time. Uranus moves very slowly through the zodiac, and yet often signals sudden action.

Many astrologers have Uranus prominently placed in their charts. This planet therefore has a remarkable influence over the individual astrologer's life. The Uranus-centered chart can reveal how you personally understand and harness the apparent abruptness of Uranus through understanding of the larger cycles involved.

The Sun transits the zodiac in one year; Uranus requires eighty-four years. This eighty-four-year cycle of Uranus has a profound effect on our human life. As it moves through the twelve signs, the energy of Uranus is shifting from one area of life to another. How can we make use of Uranian energy? What is the value of that energy to us and to our clients?

- We can examine the transpersonal map of energy in the Uranus-centered chart to help focus on the mind of Uranus at the birth time.
- There are five separate viewpoints for the planetary logos of Uranus.
- The moons suggest ritual styles or processes for achieving equilibrium. The elements the moons occupy demonstrate which alchemical processes work best for the individual.

Astrologers can clarify consulting roles through examination of Uranus and his moons. Every astrologer should understand how his or her intuition works, whether through some concrete mechanism (earth), a feeling approach (water), logical processes (air), or direct inspiration (fire). The Uranus-centered chart shows the most likely way to excite your astrological practice.

Comparison of the placements of the moons of Uranus with the placement of Uranus itself in the geocentric chart clarifies the best use of Uranian energy for any individual.

Examples

Arthur Young
Arthur Young, inventor of the Bell helicopter, documented his life with astrology in his book *Nested Time*. This book includes numerous dates for events and Young's analysis of the astrology involved. Naturally he noted the date when Bell announced the helicopter's creation, April 30, 1944. His intuition led him to consider the helicopter early in his life, and intuition drove his creative efforts all his life in scientific and other areas.

In Arthur Young's Uranus-centered birth chart, all the moons of Uranus fall outside the tight bundle of the other planets. Pluto and Neptune both are well within the bundle, so all the indica-

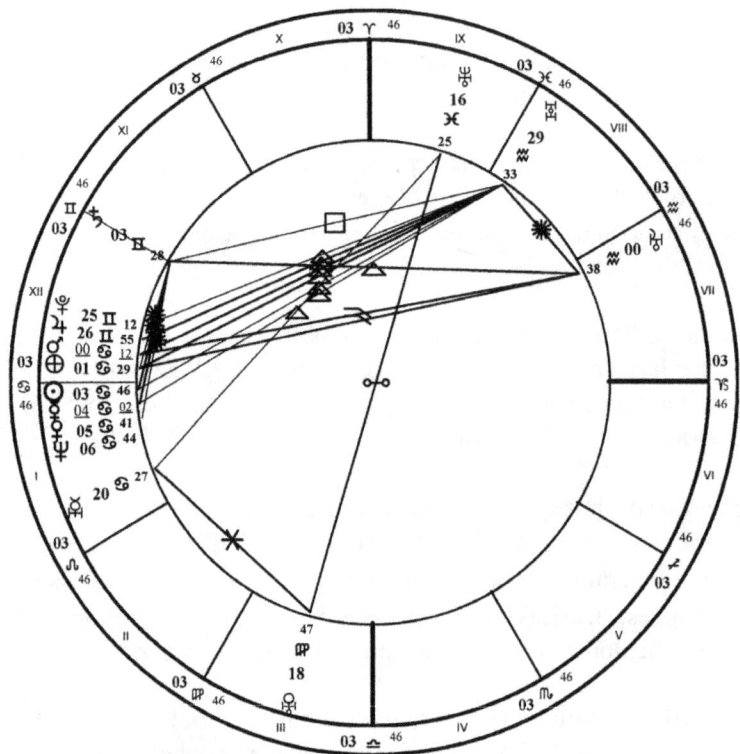

Arthur Young Uranus-Centered Birth Chart
Birth data: November 3, 1905, 10:22:34 am LMT, Paris, France

tors of Uranus's action lie outside the ordinary experience of the planets. The following are interpretations of Uranian moons in Arthur Young's birth chart:

Miranda in Aquarius: Miranda in Aquarius suggests that you see a fine point in the center that is totally, utterly and completely true, accurate, and stable. Yet you may not immediately know how to get there through uncertain territory that surrounds the center. You have the ability to see incredible detail at the center, like at the eye of a hurricane.

Ariel in Virgo: Ariel in Virgo focuses on the details of how things fit together. Your practical side gets irritated or worried when your intuition points in a different direction from your intellect. You wrap yourself in emotional insulation by keeping the facts close to you, even when the outcome is already determined. You may need to learn to let go of things you no longer need.

Umbriel in Cancer: Umbriel in Cancer indicates the importance of feelings to your well-being. You often use your intellect to figure out how to provide for others, especially children. Occasionally you will want to redirect your reasoning power to nurturing yourself. Your judgments are based on internalized feelings much of the time.

Titania in Aquarius: Titania in Aquarius reflects an emotional need for stability so you can allow your intellect to soar freely. When you cultivate a center within yourself, you feel capable of huge endeavors. In the process you keep one eye on the end result a second focus on each step of the process, and a third focus on the people who you seek to aid.

Oberon in Pisces: Oberon, the fifth moon, reflects perception. You change psychologically when rigid ideas dissolve. Early on, waves of unconscious desire sometimes overtake reason. Later, with desires satisfied, you settle down, unwinding the seemingly misdirected energies that have been at play.

In addition, aspects between the moons indicate how Uranus energy worked for Arthur Young:

Ariel Opposition Oberon: Sometimes emotional relationships become so intense that you completely lose sight of other important matters. Choose your partners carefully and build trust and security between you. Then emotions will enhance spiritual experiences.

Titania Semi-square Oberon: Even as life pulses around you, you identify key insights that drive your own actions. Use your psychic senses to find the best partnerships. You have to slow down to perceive the right choices.

Umbriel Trine Oberon: When you focus your mind through meditation, insights concerning spiritual life emerge. Spiritual insights lead to even greater moments of understanding when you see subtle connections.

From these interpretations, only a tiny sample of the whole Uranus-centered chart, the intuitive basis for Arthur Young's career is spelled out. In addition, the Uranus-centered chart reflects his astrological interests, which were strong enough for him to maintain an astrological journal throughout his life as a document for every kind of human activity, thought, and emotion.

The Brooklyn Dodgers

The Brooklyn Dodgers played their first National League game on April 19, 1890. The team provided local rivalry for the Yankees over the years, and in their first sixty-four years of existence the Dodgers didn't win the World Series. Then came the 1955 World Series. Talk about surprises. A writer for the *Washington Post* was quoted on October 5, 1955: "Please don't interrupt, because you haven't heard this one before: Brooklyn Dodgers, champions of the baseball world, honest." How like Uranus does that seem?

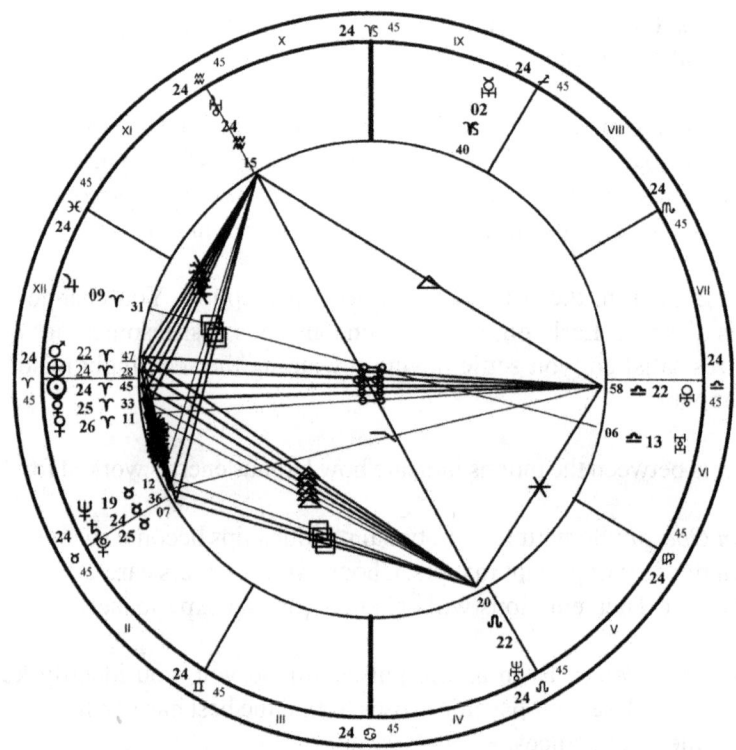

Brooklyn Dodgers First National League Game, Uranus-Centered Chart
Birth data: April 19, 1890, 1:31 pm EST, Brooklyn, New York

The Uranus-centered birth chart—the chart for the Dodgers' first game in the National League in 1890—contains a Mystic Rectangle of Titania, Ariel, Oberon, and the Mars-Earth-Sun-Mercury-Venus conjunction. Because the Mystic Rectangle includes three of Uranus' moons, we can expect the working out of Uranian energy to be magical; the 1955 series did not disappoint.

The 1955 World Series ran the full seven games. Podres pitched a no-hitter in the seventh game and also became the first pitcher to win two games in one world series (he won the third game as well). Duke Snider became the first player to score four home runs in one world series. Perhaps the most startling thing was simply winning the series. Uranus took sixty-four years to bring a win to the Dodgers and begin to restore balance to the baseball world of New York City.

Although Jackie Robinson, the first black man to play major league baseball, was a fading start by 1955, he and Branch Richey, the manager of the Dodgers who hired him, represented a couple of wild cards of the Uranian variety during the lead-up to the 1955 World Series. Johnny Podres was named most valuable player for the World Series, Branch Richey ended up in Saint

Louis, the Dodgers ended up in Los Angeles, and Jackie Robinson ended up in the baseball hall of fame. Mickey Mantle, star of the New York Yankees, only batted once during the series due to an injury. In 1958 Roy Campanella, another hall of fame star on the Dodgers team, was paralyzed in a car accident in 1958. Everything about the Dodgers has a Uranian tinge! (Roy Campanella was born November 19, 1921, at 10:45 pm in Philadelphia. Data source is birth certificate from *Contemporary American Horoscopes*, via AstroDatabank.)

The following interpretations provide a glimpse into the makeup of the Brooklyn Dodgers. The "strong" notations come from the interpretation report available with Intrepid astrological software, and mean the aspects have very close orbs.

Ariel: Ariel shows how you may best be inspired to bring about change. Here, intuition points you toward your creative promise. The fire of inspiration is evident in Ariel's placement and aspects. Creativity is a core capacity reflected in Ariel's placement.

Ariel in Libra: Your intellectual processes flourish when you allow intuitive insight to enter into your logical plan. As you comprehend the meaning of your intuition and fit it into your logical scheme, you discover a wealth of potential. Ariel in Libra suggests a serious need for objectivity. You need to appreciate the balance of waxing and waning energies.

Ariel Trine Titania (strong): Emotional fire seems great in your relationships. The best results come from well-established, trusting partnerships.

Ariel Sextile Oberon (strong): Solidity and security provide the right environment for seeking a higher spiritual level.

Sun Opposition Ariel (strong): You achieve fiery balance in partnerships.

Mercury Opposition Ariel (strong): Interactions with others are filled with emotional content.

Venus Opposition Ariel (strong): Sparks fly when you find compatible relationships.

Mercury Sextile Titania (strong): Your desire to understand your personal mission in life benefits from active communication with others. Having diverse projects and ideas helps you to further your mission in life.

Earth Sextile Titania (strong): Intelligent action, for you, is sometimes part of a wild and crazy effort when seen from anyone else's perspective. Using practical points, match up psychic or intuitive insights to reach logical conclusions.

Mars Sextile Titania (strong): It's a good thing you have strong intuition because your sometimes reckless behavior could otherwise be a problem.

Titania in Aquarius: You have a natural advantage of comprehension and ingenuity. When you cultivate a center within yourself, you feel capable of huge endeavors.

Titania Opposition Oberon (strong): You discover that many people you meet have very different values from your own.

Sun Trine Oberon (strong): Sudden changes bring new circumstances into your life.

Mercury Trine Oberon (strong): Your have a rather practical mission in life.

Venus Trine Oberon (strong): Work together with associates to manifest your ideas.

Mars Trine Oberon (strong): You have moments when your deep spiritual values direct and control your actions. When spirit is in the mix, you have great success.

Oberon reveals the nature of your perceptions from Uranus' perspective. Here you get a glimpse of the power of magic in your life. Oberon shows how you perceive the best next step in any activity. When you tune in to this energy, you often can remain a step ahead of just about everyone.

Oberon in Leo: Oberon in Leo provides the fullest potential for inspired action in the material world. Of course you thrive on meaningful results and will dump an idea that doesn't show future promise.

Summary

The Uranus-centered chart provides a transpersonal picture of the intuitive energy behind quantum change, unencumbered by individual personality. To the extent that we learn to read these charts we can appreciate the planetary logos of Uranus in terms of the creativity of this planet at specific birth times and therefore more fully understand our own creative intuitive potential.

Revelation is real. We each understand it in a slightly different way. The Uranus-centered chart can show us a lot about how we gain such deep insight. Here are some questions to ask about any Uranus-centered chart.

1. What was the direction of the planetary logos of Uranus at the time of an individual's nativity? By examining the placement of the moons of Uranus, you get a picture of what was hap-

pening for Uranus at that time, and thus what the impulse of Uranian energy is for the individual.

2. What was the effect of sign placement of each of Uranus' moons? You can consider the process of each moon to understand the whole configuration more completely.

3. What elements were occupied by each moon and what is the significance of that?

4. How do these placements reveal the best path for achieving balance and equilibrium?

5. What do the sign and element placements of the moons show about the creative potential of the individual?

6. How does the Uranus-centered chart demonstrate the truest intuition of the individual?

7. What ritual style best serves an individual in personal affairs as well as in interactions with others?

8. How does the Uranus-centered chart describe your own most effective approach to astrology and astrological counseling?

The next chapter turns to Neptune and the consideration of imagination, fantasy, and psychic powers.

Chapter Eight

Neptune-Centered Charts

Please allow yourself to take a fantasy trip as you read this chapter about Neptune. Let go of any concept of particular form and enjoy the subtle movement from words that make perfect sense to sentences that make imperfect wordings . . . resting your mind from the tension of understanding . . . perceiving that seeing can be feeling different about knowing . . . letting go. . . .

In the play by the same name, Hamlet wishes that he could let go of the reality he was experiencing—a Neptunian emotional reaction to which we all can relate:

> Oh, that this too too solid flesh would melt,
> Thaw, and resolve itself into a dew!
> . . . How weary, stale, flat, and unprofitable
> Seem to me all the uses of this world!—*Hamlet*, Act I, Scene ii

Hamlet feels despair, a wild tearing of the hair in frustration bordering on insanity. He is struggling to deal with grief; he sees his father's ghost and experiences utter powerlessness. When the astrologer speaks of Neptune, disillusionment and despair can enter the conversation as companions we have known in our journeys through life.

Discussion of disillusionment brings to mind the child who asked for a pony and got a plastic Trigger; it feels like the adult who wanted Haagen Daz ice cream and got plain tofu. Nothing can match the flatness of those moments when you realize that you cannot have what you thought you might have wanted, and that, further, you are not certain what it is that you wanted then or may desire in the future, if you could only remember what was happening when you thought that idea then. . . .

Neptune and Associated Mythology

Etruscan Nethuns began his existence as the god of wells. Later his role expanded to god of all water, including the ocean. He emerged as a strong influence on Neptune, the Roman god of water and the sea who in turn shares qualities with the Greek Poseidon. In addition to ruling water, particularly the ocean, Neptune and Poseidon were worshipped as the gods of horses and sometimes earthquakes.

Among the major gods of the Greeks, Poseidon nevertheless remained in the water most of the time. He governed both fresh and salt water. Poseidon's days were filled with conflicts with Athena, Odysseus, King Minos, and others. The turmoil of the sea reflected his apparent emotional turmoil.

Another ocean deity is Tiamat, the Babylonian goddess of the waters. Chaos is an integral part of her being. Sometimes perceived as a dragon or serpent, she represents two distinct kinds of water. First, there is the water that covers the planet. Second, in Tiamat's world there is also water occupying the space we generally reserve for air. This second type of water relates to spritual matters. Both types of water stimulate practical and imaginative creativity.

With Tiamat, you can experience intense delusion and disillusionment, or you can experience profound mysticism and ecstasy. Both irrational and transcendant, according to Raven Kaldera, Tiamat reflects the astrological range of Neptunian energy.

Norse myths tell of Aegir, a Germanic sea god, and his wife Ran. Aegir was a very old man with white hair who rose from the depths to destroy ships. Ran caught sailors in her nets and drowned them. Thor supposedly forced Aegir to supply ale to the gods.

The Orisha Yemoja not only governs all water, from amniotic fluid to the ocean, she represents the love principle. Maternal and nurturing, Yemoja is mother of the universe and the visible part of the ocean (she rules as Olokun in the ocean depths). She can be very temperamental, yet she is also quite protective. She is a sorceress of great power.

If Neptune's depictions in various mythologies are confusing, that reflects the astrological nature of this planet nicely. The mental range of Neptune's expression equals the emotional. Neptune spans the scale from deception and illusion to a sensitive inspiration from which art works are made. The high level of wisdom goes beyond mere information and logic, entering levels of experience often restricted to drug-induced ecstasy. We often think of Neptunian processes as magical, primarily because we don't understand them.

The mind needs to move through ideas instead of becoming mired in them. As movement begins, the Self can emerge from such a wandering, rambling, wavering sense of what is really important. We can learn without structure if we open our minds and hearts to the Infinite. Visualize what you want and it can be yours. Magic carpet or personal effort? Either will do nicely.

The alchemical process of Neptune dissolves hardened structures from the past in order to make way for a new, more complete or useful structure. The alchemists knew that dissolving a metal is necessary in order to purify it. They utilized a number of processes to dissolve compounds. The comparable stage of personal process demands a great deal of courage and curiosity because it is frightening to experience the dissolution of the old mental structures before new ones are in place. There can be a sensation of walking into quicksand.

Neptune represents the energy released during the alchemical process of reduction to the *prima materia*. This energy can take an obvious form, such as heat, or it can take the form of psychic libido. As you purify your experience, cleansing it of any emotional residue that has become attached to it, you eventually find at the core—the Self. This Self is not the entire being, but rather an essential part of us through which you experience greater wholeness.

Tension exists between the two possibilities. You can be so soft that you lose touch with your senses altogether, or you can be so directed and rational that you cannot make a decision from the heart. Neptune controls a broad continuum from the very best you can perceive to the very worst. You have to achieve your own balance between its severity and mercy.

People always want results; with Neptune, however, results are not as important as process, at least initially. We do eventually hope to see results, but patience is a large part of the dissolving process of this planet. The Lord moves in mysterious ways. The mystery itself consists in the archetypal energy of Neptune. While Saturn pushes us to analyze and objectify experience, Neptune allows us to simply experience without labeling or conceptualization. Yet mysteries are revealed from time to time, providing us with at least an occasional glimpse of personal mission and encouraging us to try to fulfill that mission. Misunderstood energy can produce misdirected mission—the appearance of evil. Yet we know that the evil is our own conception and not the nature of archetypal reality.

In Shakespeare's dramas, Neptune provides both pitfalls and solutions. Hamlet opts for obtaining solid evidence before he acts, and waiting is his downfall. It seems we must be ready for transformation when it is ripe or the decoction may sour somehow. Neptune may not deliver all that it promises.

Neptune Key Words

Physical	Emotional	Mental	Spiritual
Pineal gland	Receptive	Imagination	Compassion
Solar plexus	Sympathy	Fantasy	Mysticism
Body functions	Fear	Genius	Inspiration
Deformity	Gullibility	Reverie	Metaphysics
Relaxation	Unconscious	Telepathy	Divination
Drugs			Psychism

Neptune-Centered Charts

Neptune has two major satellites, Triton and Nereid, and at least six minor ones. Mars and Neptune share sixth ray energy in esoteric astrology; their moons could hardly be more different in appearance and expression.

To begin with size, Mars is 6787 km diameter, Phobos is 20x 23 x 28 km, Deimos is 10 x 12 x 16 km. Neptune's diameter is 49,500 km, Triton's is 3500 km (over half the size of Mars), and Nereid is 400 km. Both Phobos and Deimos are tiny, even compared to a relatively small Mars; Triton is about 1/14 the diameter of Neptune while Nereid is about 1/124.

Orbital periods around their respective planets are very different as well. Phobos is racing around Mars in a few hours and Deimos is a bit over one day. Triton has a period of nearly six days; Nereid's period is 562 days. Nereid is as slow as Phobos is fast.

Triton is one of the few satellites that follows retrograde motion around the planet. Nereid has a very eccentric orbit, more elliptical than many comets, with a distance from Neptune varying between 140,000 and 9,500,000 km. Yet this unusual moon orbits Neptune, a planet with a very circular orbit. The combination can help to explain our astrological understanding of Neptune.

Mythology of Neptune's Moons

Triton and Nereid (actually there were numerous Nereids) were grandchildren of Oceanus. Triton's parents were Poseidon and Amphitrite, while Nereid was born to Nereus and Doris. Originally a Libyan deity, Triton was half-man and half-fish. He was capable of raising or quieting the waves and was responsible for saving the argonauts from destruction. He had the gift of prophecy.

Nereid, considered propitious to sailors, spent her time spinning and singing in Oceanus's golden palace. She is the compassionate witness of drama and represents the mystery of sea life.

In mythological references, Phobos and Deimos represent the fierce qualities of Mars that can strike terror in one's heart. Yet they are tiny moons with relatively little to recommend them. Massive Triton and Nereid, by comparison, embody rather elusive qualities such as prophecy and compassion.

I find myself comparing these qualities and giving them their relative weights in relation to the size of the particular moons. Can it be that human potential for prophecy and compassion far outweigh any fear, anger, or terror? Astrologers foresee events clearly in many cases. While I attribute this skill primarily to the ability to add two and two astrologically and to project the probability of certain outcomes, I also believe psychic abilities help us all.

Neptune's moons can be compared by relating the qualities of prophecy and compassion. To take prophecy first: Triton rotates around Neptune in retrograde movement. This suggests a parallel to many of the teachings of transpersonal psychology, and particularly of eastern religions. It is my belief, supported by many religions, that each of us is a fully realized being who has forgotten the fact of that realization. We seek lost understanding as though we never had it at all. Seeking is based on our limited human perceptions, limited by virtue of the fact that we find ourselves in relatively slow material bodies. We must perceive polarities in order to perceive the phenomenal world.

Prophecy may be largely a matter of remembering an already fulfilled future. It is rather like watching a movie for a second or third time. We already know on some level what will happen, even if we have forgotten the details; as the movie unfolds we remember each scene. Like déjà vu experiences, we suddenly become aware that we have done this before. In each case there is the profound sense that we are not seeing or doing something new, but re-doing or perceiving again.

Astrologically, the prophetic experience relates to retrograde motion. When planets are retrograde we get to re-think, re-experience the aspects made by the planet while it was direct in motion. We know that the future has not yet occurred, and we know that we cannot logically expect our minds to traverse time to see the future. Yet we do have experiences in which we have become aware of the future.

The medium of water where mythical Triton lives has the eeriness of shifting light and shadow. We cannot be precisely sure of what we are seeing. At the swimming pool or even in the bath tub we become aware of the bending of light as it enters the water, distorting shapes. Combine this with the sensation of remembering something that has not yet happened and we get a measure of the confusion prophecy causes. The illusionary quality of Neptune results from the shifting base of reference.

Through the practice of meditation you can master confusion arising from Neptunian thought processes. As you quiet the mind, you filter out the interference of moment-to-moment thoughts, clearing a path for messages to follow. As you meditate, don't seek anything in particular, but gradually become aware of significant themes. Each person has a unique set of psychic skills, and meditation is one way to identify and develop those skills.

Nereid and Compassion
Nereid is the holder of compassion. This moon, in its eccentricity, shows you the way to empower compassion in your life. You don't need to be logical or consistent about compassion. You don't need to reserve it for truly deserving individuals. You don't need to have your compassion under control. You can let it fly out as far and as fast as it wants; you can perceive its return. As you soften to the effects of compassion, you become even more capable of it.

Yet you can also exercise intelligence in compassion. You learn when true kindness has a sharp edge. For example, is it compassionate to reprimand a child for running in the street? How can you best teach this lesson? Is it compassionate to provide medicine to relieve pain when you can? How can the principles of compassion be used in the training of employees? Everywhere you look, you find occasions for the application of compassion and for the study of its finer points.

The more you understand compassion, the further you may seem to move from what logic tells you. Compassion has its own internal logic, independent of what you supposedly "know" about the world. Nereid has her own internal timetable for her orbit around Neptune. She leaves the very orderly circular orbital demands to the planet while she takes excursions outside the proper plane of things. You can leave the ordinary realm of thoughts and objects to exercise as much compassion as you can tolerate. I contend that we all become better people by so doing.

Examples

Timothy Leary
One of the gurus of the psychedelic movement, Timothy Leary encouraged a whole generation to drop out so they could find their power in new ways. He fearlessly researched LSD and other drugs and administered them to Allen Ginsberg and other poets of the time. Leary had his own

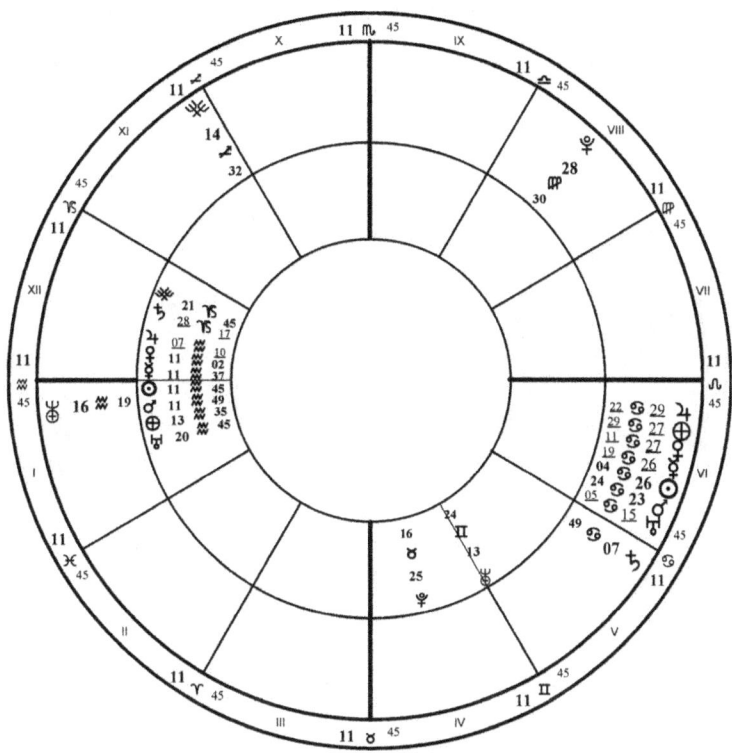

Timothy Leary Neptune-Centered Chart (inner) and Death (outer)
Birth data: October 22, 1920, 10:45 am EDT, Springfield, Massachusetts;
Death data: May 31, 1996, 12:45 am EDT

problems both while a student and later as a professor. In 1963 he was relieved of his teaching position at Harvard, presumably for missing classes. He was arrested for possession of marijuana in 1965 and again in 1968. He escaped from a prison farm and left the United States, but was later returned and did a short term at Folsom Prison and was released April 21, 1976.

After he was released, Leary espoused space exploration. He thought the Internet was the drug of the 1990s. He mingled with religious groups and at one point declared himself a pagan. He had his death videotaped, passing on at 12:45 am, May 31, 1996. A sometime fan of cryogenics, Leary was ultimately cremated and part of his ashes were sent into space on April 21, 1997 (lift-off at 11:59:00 am EST).

In the Neptune-centered birth chart, two aspects define Leary's Neptunian attitude toward drugs and society. First, Pluto forms a trine to the midpoint between Saturn and Triton. Second, Nereid forms a trine to the Earth. The following interpretations describe the configuration:

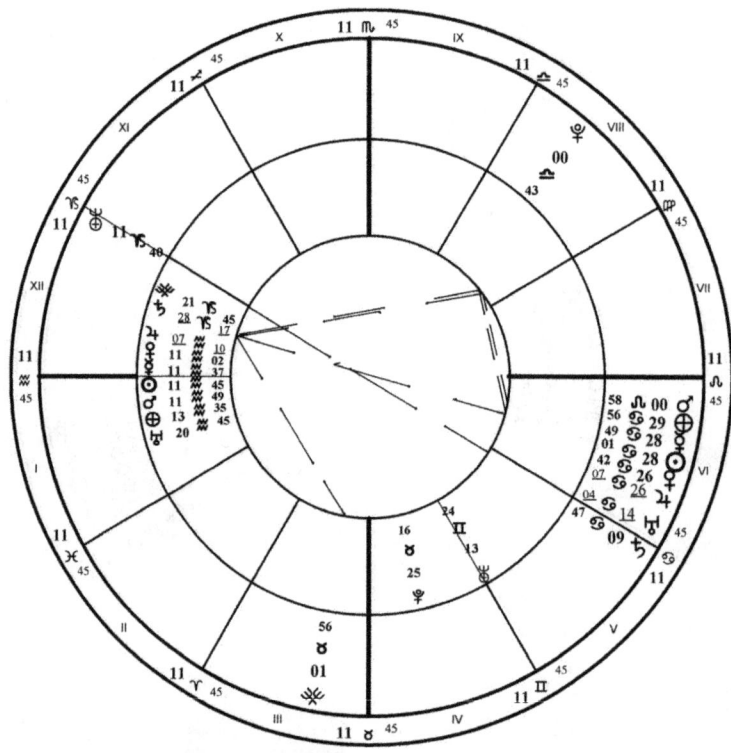

Timothy Leary Neptune-Centered Chart (inner) and Lift-off into Space (outer)
Lift-off data: April 21, 1997, 11:59 am EDT

Saturn Conjunct Triton: Prophetic wisdom—a profound intelligence concerning the future—is something you develop by aligning your intuition with concrete facts so that both make sense in a single context.

Saturn Trine Pluto: You experience cruelty in your life that must be offset by other factors. You have the potential to grow in spiritual awareness if you can avoid purely egoistic aims.

Pluto Trine Triton: Prophetic transformation is possible in your life. This sort of transformation occurs before the future events you foresee, and allows you to engage without ego interference when the times demand it.

Earth Trine Nereid: You have a natural ability to associate psychic insights with more traditional reference points in your life. Intelligent compassion means helping when you know what to do, not when you feel you should do something. Use your discriminating taste to evaluate situations.

In the Neptune-centered biwheel charts for Leary's death and for the rocket launch of his ashes, there is a very close opposition from the transiting Sun-Mercury-Venus-Earth conjunction to Saturn in his birth chart with a sextile and trine to Pluto (in Virgo in 1996, and in Libra in 1997). These identical configurations are possible because all of the planets move very slowly from Neptune's perspective. Key differences between these two dates include transiting Nereid conjunct the Earth/Uranus midpoint in the 1996 chart, whereas Nereid was semi-sextile Venus, Mercury, Sun, and Mars in the 1997 chart.

Looking forward to the next chapter, the Pluto-centered chart for Leary's death has an opposition of Neptune to Jupiter, with a sextile and trine to Uranus. The 1997 Mercury-centered chart has an opposition of Jupiter and Neptune with a sextile and a trine to Saturn. Leary definitely covered a lot of bases! His sense of timing appears to have been quite good.

Jonas Salk
Possibly the single most significant advancement in medicine in my lifetime involves the development of a vaccine for polio. Maybe my perspective is colored by two facts. First, two elder family members contracted polio, one as an infant and one as an adult. Second, my brother was in the second grade at the time of the initial tests, so he was in the test group. That year the number of children who contracted polio dropped like a stone. That was 1953 and 1954. Although Salk never received a Nobel Prize for his work, he is revered in the hearts of parents everywhere because he saved my generation and succeeding generations from a disease that crippled and killed so many. Later in his life he worked diligently to advance AIDS research.

The announcement of successful trials with the vaccine came on April 12, 1955, at 10:20 am in Ann Arbor, Michigan. Unlike many medical researchers, Dr. Salk refused potential income from the vaccine.

The more distant a planet is from the Sun, the more the inner planets tend to bunch up around the Sun. We have seen that effect in previous chapters. Now we see it even more. Although the moons and Pluto can potentially be found anywhere in the Neptune-centered chart, the only objects outside the solar bundle in Salk's chart are Nereid conjunct Pluto, with Pluto forming a trine to Mars and Triton.

In the transiting chart, Pluto is quincunx Jupiter and Uranus, the Uranus aspect having an orb of only 5' of arc. Nereid is trine the Earth in the birth chart, with an orb of only 2', and the Sun is square Triton with the same orb. These tiny, tiny orbs reveal the timing potential of the moons of Neptune. Additional aspects include Uranus trine the Sun, and transiting Sun square Mars. Transiting Triton was conjunct Nereid at 10:55, just 35 minutes after the announcement, suggesting that the impact of the news was already hitting Dr. Salk in ways he may not have fully anticipated (he never achieved the full support of his colleagues).

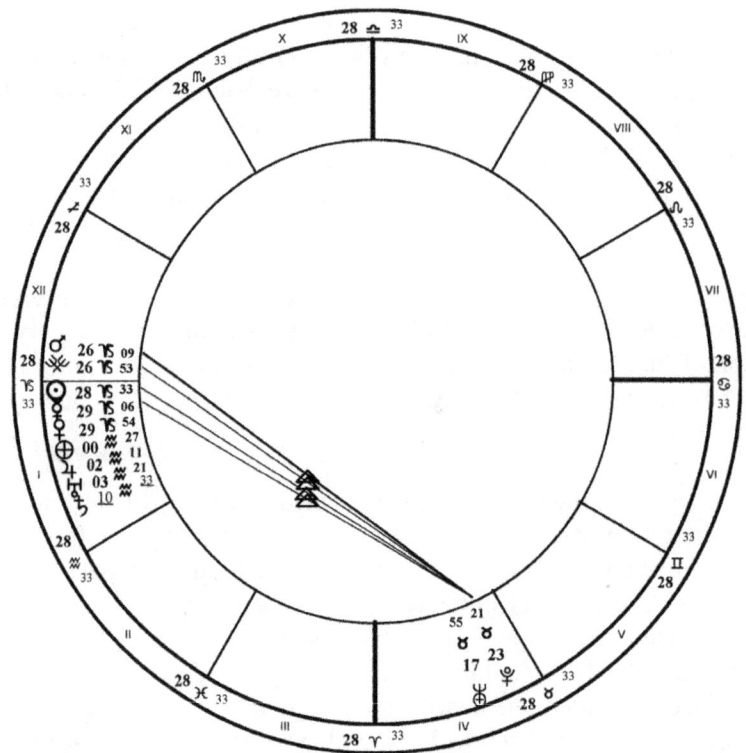

Jonas Salk Neptune-Centered Chart
Birth data: October 28, 1914, 7:30 am EST, New York, New York;

Particularly significant are transiting Mars square Mercury with an 11′ orb, while transiting Mercury is square Mars, also an 11′ orb. These mutual aspects reinforce the message of the chart—the successful vaccine was a monumental step forward in medicine. Another pair is transiting Mars square Venus and transiting Venus square Mars. Many a mom and dad fell in love with the man who saved their children from the fear of polio. Keep in mind that with the bundling effect in the planet-centered charts of the outer planes, these mutual aspects are not so rare.

The following transit-to-birth interpretations are unique to the Neptune-centered chart. Other aspects represent familiar possibilities in the tropical chart, and their meanings remain the same with the caveat that they be considered from Neptune's perspective—including devotion to one's mission and work.

Sun Square Triton: Your powers of prophecy build upon a foundation of love and remove the veil of illusion from the world. The challenge is to present your understanding of the future in ways that encourage others to believe you.

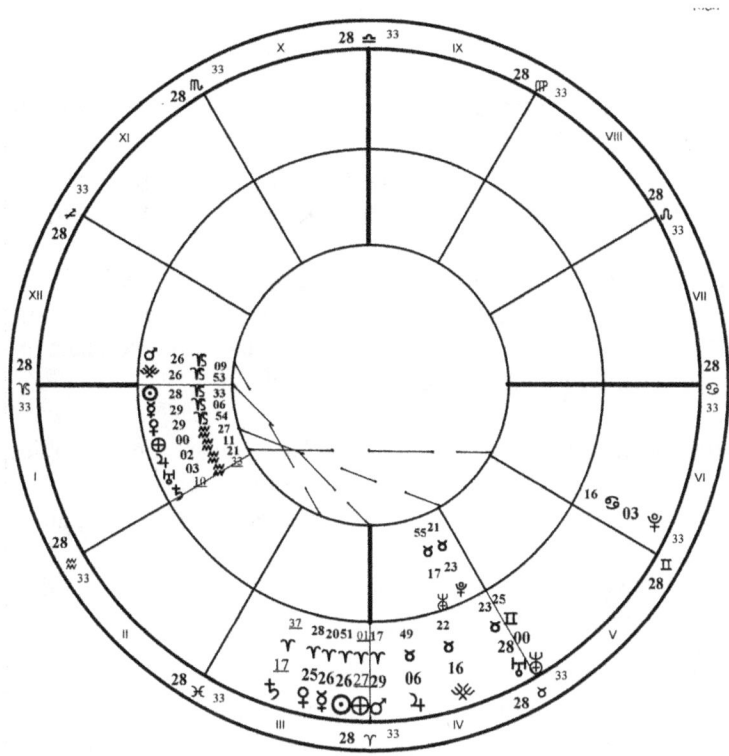

*Jonas Salk, Neptune-Centered Chart (inner) and Polio Vaccine Announcement (outer)
Announcement data: April 12, 1955, 10:20 am EST, Ann Arbor, Michigan*

Mercury Square Triton: You have the gift of prophetic speech. Sometimes you have to work very hard to get your unique vision across to others.

Triton Conjunct Nereid: Neptune's power is concentrated within your consciousness. Prophetic talent joins with compassionate impulses to guide your action in relationships. Devote your mind to spiritual values and your talents will grow exponentially.

Nereid Trine Earth: You have a natural ability to associate psychic insights with more traditional reference points in your life. Intelligent compassion means helping when you know what to do, not when you feel you should do something. Use your discriminating tastes to evaluate situations.

Summary

This chapter on Neptune has been the hardest to write of all the planets, probably because of the nature of Neptune. That said, the Neptune-centered chart is a viable tool for expanding upon the nature of Neptune in the geocentric birth chart and for clarifying or pinpointing events with profound precision. It also allows the astrologer to assess the range of psychic powers for any individual.

Because Neptune is so far from the Sun, all the planets tend to bundle closely together, with even Jupiter only a few degrees away. Pluto can be found anywhere. For example, in 1891 AD, Neptune was conjunct Pluto. This means that in the Neptune-centered chart the Earth, Sun and inner planets were opposite Pluto in the sky.

For everyone currently alive, however, the moons of Neptune are the only objects that can form an opposition to anything. Thus the location of the moons is critical in the Neptune-centered chart as delineators of Neptune's energies of psychic talent, compassion, and so on.

1. Where are Triton and Nereid in your Neptune-centered charts? Consider the elements, signs, and aspects they form, especially to each other.

2. Consider the kind of prophecy you may have or become adept with. The element of Triton can describe the part of the phenomenal world you can contact most easily. The sign suggests the means you use or the medium through which you reveal your knowledge.

3. Consider how you tend to express compassion. The sign and element show your natural tendency in this regard.

4. What sign does the Earth occupy? This tells how your talents are viewed from the Neptunian perspective.

5. How can compassion and the use of psychic talents moderate or modify your experience of Neptune (as seen in your geocentric chart? How can they help clear up misunderstandings, for example?

6. What happens to fear when you utilize Triton and Nereid (psychic talents and compassion) in your daily life?

7. Consider using the Neptune-centered chart (or any planet-centered chart) as a visual focus for meditation. Let the shape of the chart speak to you about your own compassion and psychic ability.

The next chapter looks at Pluto and his moon Charon. Here we find possibly the most bizarre pairing in our solar system!

Chapter Nine

Pluto-Centered Astrology

Which of the archetypal expressions relates to Pluto? The power of Pluto can be seen on the Fool of the Tarot and the Trickster of mythology and Jungian psychology. In his book *MythAstrology*, Raven Kaldera lists these archetypes for Pluto: Destroyer, Rebirther, Pain Giver, Power Seeker, Purifier, Death god or goddess, Guardian at the Gate of Mystery, and Keeper of the Depths.[24] Pluto's mythology reveals some dark, grim characters who govern many of life's deepest mysteries.

Mythology of Pluto

When Zeus and his brothers Poseidon and Hades divided up the known universe, Hades got the underworld. Reportedly somewhat shy, this suited him well. Never the satanic governor of the Christian Hell, Hades nevertheless was grim. To visit his realm required that you consume nothing lest you be compelled to remain in the Underworld forever.

Yet Hades wanted a consort and stopped at nothing to obtain one. Various stories tell of how he raped Kore or captivated Persephone, taking his love back home with him. However, once Kore/Persephone entered the Underworld, she began her own transformation, a change we can each experience as we begin to deal with our innermost problems and face the crises of their resolution. Hypnos ("Sleep") is another dweller in the Underworld who never sees the light of the sun. His brother Thanatos ("Death") and his son Morpheus, the god of dreams, are also associated with Hades.

Egyptian Osiris has traits in common with Pluto. However, when he became ruler of the Underworld, he was generally seen as kinder and more helpful, much as he was when he was Pharaoh. He reportedly conquered nations peacefully, using slow, steady effort. Slowness is a quality of transformation—you have to handle each step of the process with care, and force usually doesn't accomplish much. This reflects the reality of the planet Pluto—its slow repetitive passage through degrees of the zodiac permits you to refine your sensitivity and wisdom.

Turning to feminine depictions of Pluto, we find some truly intense, grim goddesses. Durga, a female Hindu deity, provides the strength of personality you need to become strong. Although she is seen smiling and confident, she does not deny the problems you encounter. She fiercely pursues demons, making them her first priority, just as Pluto provides the will and power to face your personal transformation. Kali dressed in the bones of her victims and even devours her consort in some depictions. Hel, a member of the Norse pantheon, is so terrible to look upon that her name was given to the Christian underworld. This goddess, however, treats everyone equally in her realm.

The Tarot Fool, another Pluto figure, appears to be reckless beyond his skill. Or is he? He appears to step off the edge of a cliff into the void without any conscious concern. We identify this tendency in ourselves to ignore warnings and to move into perilous new experiences without preparation, and we know we can be hurt. Yet the Fool may also be happy-go-lucky because he has developed his inner resources and these powers carry him through situations fraught with peril. If he has pursued the path of individuation, the alchemical process of refinement, and if he has a grip on his relationship to his instincts, then he can safely avoid trouble. Pluto emerges as the power behind both of these expressions of the Fool's potential.

Interpreting Pluto

When Pluto becomes active in your life, you may at first feel like you have been metaphorically raped. No gentle onset here! Pluto is in your face from the get-go. Interestingly, Pluto experiences focus on hidden, unconscious issues that arise from deep within you. Thus you are the source of your own Pluto conflicts in many cases. Not everyone is coerced by others—we also have our own dark secrets to deal with. Many astrological clients experience difficulty with Pluto aspects because they are terribly out of touch with some part of the psyche. When you ignore or suppress one side of your nature, you inevitably are called to task for it. You will find, on the other hand, that Pluto is a powerful and constructive force when you consciously relate to your inner power and listen to instinctual messages, bringing them into consciousness.

The aspects of Pluto show how an individual customarily approaches the use of power. When Pluto aspects other planets or the angles, you can identify areas of life where power is most important through the house placement of Pluto and the other planets. Further, you can perceive the methods of focusing personal power that will be most easily and dramatically available for the native. The aspect itself will describe the "how" of the process; you can hypothesize about a person's ability to use power and you can also predict the constructive or destructive tendency, depending on the person's strength of personal will. When Pluto is involved in an aspect you see how the native connects most directly with the tremendous power of this planet, both on a conscious level and in unconscious patterns.

The placement of Pluto in a chart and its aspects provide a map of the potential utilization of power for the individual. This map influences the person only to the extent that power is available; the astrologer can utilize the map to assist the development of greater awareness of the potentials involved.

In terms of intrinsic health, Pluto pinpoints the positive intention of every action in the world. At first Pluto and its aspects may reveal difficulties. The unconscious expression of Pluto in aspect to other points in a chart can indeed seem totally out of your personal control and therefore will be perceived as a problem that has no solution.

The less conscious expression of Pluto is identified in psychological literature as obsessive thought and compulsive behavior. Obsessive thought occurs if you have only one way of thinking about a set of circumstances; that is, you have unchanging beliefs. Compulsive behavior is the resultant activity. You have a set of beliefs about what will work and you try to support this belief by utilizing the same behavior again and again, regardless of whether it is actually useful. At the very least you derive some comfort from the repetitive nature of your actions and you recognize the behavior as something you know how to do.

However, two problems can arise from this pattern of behavior and belief:

1. You may find that the compulsive behavior now causes pain. If you wash your hands too often, for example, the skin may chap and crack and become sore. Now there is a bind: do you continue to perform this act which offers comfort on an emotional level even though it creates physical pain?

2. You may discover that you want to pursue activities that are inconsistent with your compulsive behavior. For example, an athlete may want to ski the high basin, and he may have the skill for this activity, but his compulsive avoidance of high places interferes with this desire.

In both cases change is possible, yet the inner Fool/Shadow is holding firmly to the old behavior to satisfy emotional needs. People all need to change, but often we don't know how. However, if we look at the problem as an expression of an inner urge to security and satisfaction, then we evaluate the discomfort of the activity in a different way. The Fool within is either unconsciously moving through life, recklessly ignoring potential pitfalls until he is mired in them, or he is fully enlightened and thus able to walk unconcernedly where angels fear to tread because of his direct and conscious connection with the instinctual self. Each person is somewhere along the path of development from totally unconscious fool to fully aware and functioning being. In any case, each problem tackled is an expression of powerful effort at self-protection and self-satisfaction.

Pluto reveals positive intention for each of us. I believe that everything a person does is the best he or she is capable of. If actions seem bad or weak, then the person apparently fails to communicate with the best inner guidance. So, if you perceive that your actions are not working well, you first need to determine what the intention is.

As soon as you have grasped the less conscious positive intention, you immediately begin to develop new and better choices for yourself. Now the personal power indicated by Pluto is free to move in several directions instead of being limited to an outworn behavior. Through a process of self-study and choice development you have moved unconscious Plutonian power controlling your life into a new state in which Pluto energy functions more freely and more powerfully. You had to examine your motives, find the innate positive intention in order to free up power that was there all the time.

Pluto Key Words

Physical	*Emotional*	*Mental*	*Spiritual*
Death	Masochism	Amnesia	Regeneration
Genitals	Will to power	Rational will	Reincarnation
Convulsions		Healing	Transformation
Birth			Providence

Pluto's Process

We began a psychological journey with a discussion of Saturn and structure. Through examination of the alchemical process, we found that Uranus indicates the shattering of old structures in order to make room for something new. Next we underwent a process of dissolving and distilling, governed by Neptune. This stage involves a refining of the *prima materia*, so much so that we may lose sight of both where we started and what we wish to become.

Pluto represents the completion of the work, when we see the refined material in its brilliance. The metal is not you; it is a metaphorical part of you that you now see in a different light. You become aware that within the confusion and limitation of your old structure there existed a fine, refinable essence, and now you can perceive that essence more directly. The freed-up power of Pluto can now serve you in all areas of your life and it will also allow you to pursue further refinement of your being.

How does this further refinement occur? It can occur only if you, like a 13th century king, keep your Fool around where you can see him. You need to be aware of the Trickster/Shadow at work in your life and laugh at yourself and him when you are challenged to change once more. Pluto is only dangerous when you are out of touch with the instinctual side of your nature. Power flows when you consciously attend to the message from the unconscious. The non-rational can bring joy and excitement to you if we are able to let it into consciousness.

Pluto-Centered Charts

Of the planets with moons, Earth and Pluto are at the extreme. Earth's moon can create total eclipses on the surface. It follows the zodiacal ecliptic at a rate of approximately thirteen series per year. Earth's moon is considered to be an equal partner, governing night as the Sun rules the day. It is apparently the same size as the Sun, hence the possibility of total eclipses. Yet the Moon is portrayed as the opposite of the Earth in many respects; certainly this is true in astrology. Earth's moon rises and sets, and its changing phases signal rhythms of life on Earth. We see only one face of the Moon because the Moon rotates once for each revolution around the Earth. Major holy days are measured by the Moon's phases in relation to the Sun's position in the heavens.

On Pluto things are very different. First, Charon is over half the diameter of Pluto. Charon is more like a companion than like a moon. Charon will create an eclipse on Pluto only once in approximately 124 years. The orbital period of Charon and the revolution of Pluto are identical, making Charon a fixed object in the sky over one part of the planet. Thus Charon's apparent travel through the heavens is comparable to the change in the Midheaven or Ascendant on Earth: as Pluto turns, Charon follows that movement precisely. Pluto and Charon have no independent lives, but move together.

Diurnal motion on Earth places Pluto in a house, depending on the distance between Pluto and the Sun in the zodiac, and the time of day. The position of Charon in a Pluto-centered chart depends only on the time of day from a Pluto perspective. On a given place on Pluto's surface, Charon will always have the same position in the local sky. One might need to move around on Pluto's surface to ensure seeing the moon at all. While a point on Pluto's surface bears a constant relationship to Charon, the position relative to the zodiac is changing at the rate of 360 degrees per Pluto "day," or every 6 days 9 hours 17 minutes.

Just as the Sun and Moon provide two very different reference points concerning human personality, Charon will give a Pluto-centered description of the nature of Pluto. The angle between the Sun and Charon in such a chart describes the nature of a power partnership from Pluto's perspective at the birth time.

Charon Personalizes Pluto

The zodiacal position of Charon enriches the meaning of Pluto, a planet that can take years to transit through one zodiacal sign. With a revolutionary period of just under 6.5 days, Charon transits the zodiac nearly five times a month, whereas Pluto will move very little in that time period.

Thus while your perception of Pluto in a sign is shared by thousands of people born close to the same date, your Pluto-centric chart will reveal a unique placement of Charon, thereby illuminating your personal use of Pluto's energy in your life.

The mythology of Charon and his relationship to the god Hades (Pluto) is consistent with the astronomical relationship. The mythological Charon resides in Hades and ferries people across the river Acheron (also sometimes described as a swamp). The son of Erebus, Charon is an aged, dirty man. He requires a coin from each person who is entering Hades. He also would not allow people into his boat if they had not been properly buried. Thus we see that Charon is very much the gatekeeper of the Underworld. Charon is the servant of Pluto's bidding, and not an oppositional force in any respect.

In the Pluto-centered chart, then, the position of Charon provides a complementary interpretation of the role of Pluto in the geocentric chart. For example, if Pluto is in Virgo in every chart for a period of years, then Charon will add 360 degrees of variation to that Virgo position. It will not change the interpretation of Pluto's position, only add flavor to it. The degree meanings in Intrepid software interpretations enrich understanding of the meaning of Charon, as they do all the planets and moons.

Pluto and Esoteric Astrology

Although Pluto was newly discovered at the time Alice Bailey did her writing, she had some relevant comments. She stated that Pluto drags up all that we may not want to look at within ourselves so that we may destroy hindrances to further progress.[25] Jeffrey Green had similar thoughts on Pluto[26]. He stated that "Pluto correlates to the deepest emotional security patterns in all of us. . . . These deep-seated security needs drive us to approach certain areas in life in exactly the same way over and over again."

Bailey associated Jupiter and Pluto as rulers of Pisces (Jupiter being the traditional ruler and Pluto being the esoteric ruler, in her view). This close association gives rise to the psychic sensi-

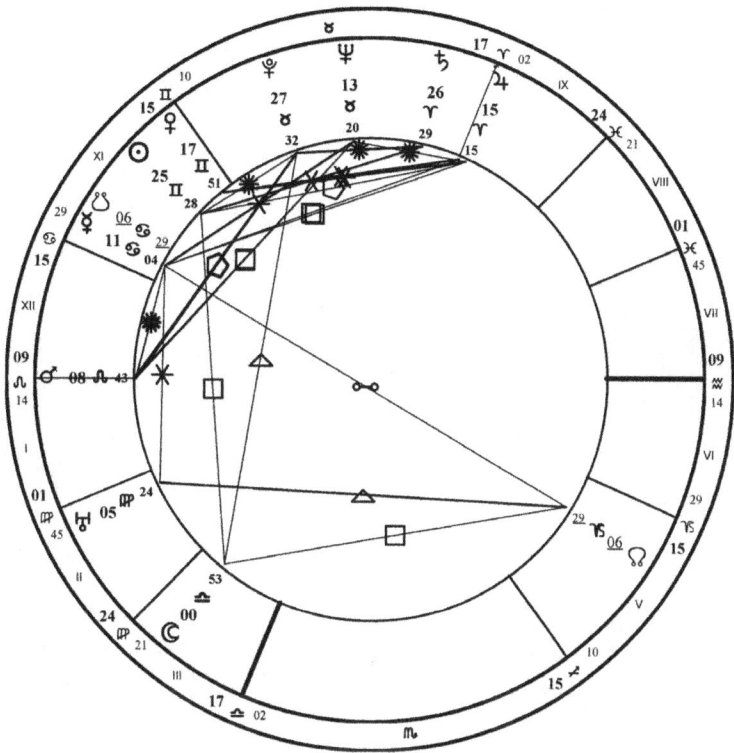

Alice Bailey, Geocentric Birth Chart
Birth data: June 16, 1880, 7:32 am GMT, Manchester, England

tivity that is often prominent where Pisces is emphasized in the horoscope. Jupiter and Pluto together reflect the lure of the evolutionary path. They further indicate processes of transmutation and eventual death of the physical body.[27] Pluto also reflects the death of desire and the death of the personality that keeps us connected to the world of opposites. For the person who has no aspirations to evolve spiritually, Pluto may express only in terms of destruction. However, both Bailey and Green make the point that once on the path, each destructive moment or event gives rise to new, creative potential, much as the Phoenix burns up and then rises from its own ashes.

Examples

Alice Bailey

We don't have a definitive birth time for Alice Bailey. Using Rudhyar's suggested birth time, Pluto and Mars form a very tight quintile and Pluto is also quintile the Ascendant. The semi-square between Jupiter and Pluto may reflect her insight into the interplay of these two planets in the evolutionary process.

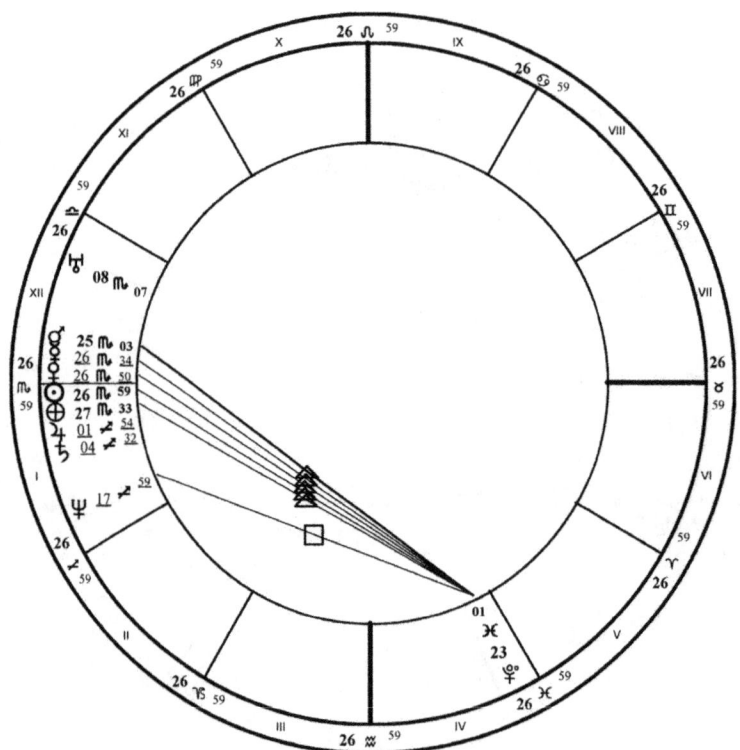

Alice Bailey, Pluto-Centered Birth Chart

Looking at Bailey's Pluto-centered chart, we see the characteristic tight grouping of planets around the Sun, a result of the perspective from distant Pluto. The only object outside the group is Charon, Pluto's moon. Charon trines Mars almost exactly, reinforcing the Mars-Pluto quintile in the geocentric birth chart and emphasizing the energetic devotion Bailey had to her work. Charon is sesquisqaure Uranus, reinforcing the Sun-Uranus quintile in the birth chart and emphasizing intuitive processes.

With Charon in Pisces in the Pluto-centered birth chart, Bailey's Pluto-centered chart reflects the importance of both Jupiter and Pluto as rulers of this sign. Charon here suggests that Pluto's energy is backed up by the companion moon's sympathetic wisdom and power.

With the Sun in Scorpio in her Pluto-centered chart, Alice Bailey belonged to a generation of people whose challenge was to harness power and will to effect transformation in the world. Bailey, through her source Djwal Khul, explains the spiritual, esoteric workings of energy in our lives, the process of spiritual evolution, and ways we can utilize power and will for our higher good. Charon is sextile the Sun, Venus, Mercury, and Mars, revealing that her many op-

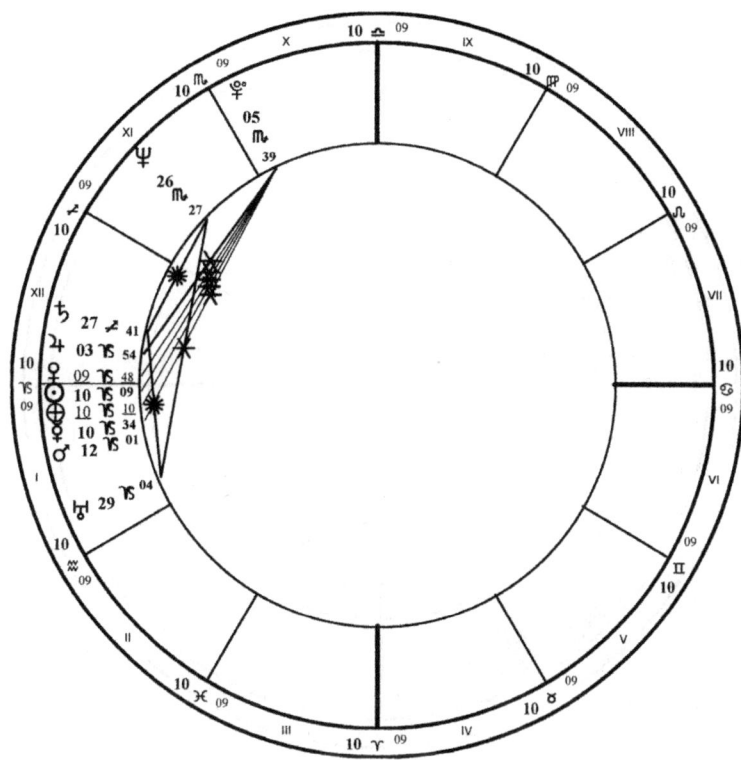

Douglas Baker Pluto-Centered Birth Chart
Birth data: December 31, 1922, 11:55 pm GMT, London, England

portunities to express power and will were not lost on Bailey, who was able to attract students, associates, and funding to continue her work after her death.

Douglas Baker
Dr. Douglas Baker was born more than forty years after Alice Bailey. He studied her work and published some beautiful volumes about esoteric astrology. In his birth chart Pluto is prominent at the Midheaven and part of a Grand Trine including Jupiter, Uranus, and Mars, with the Sun at the point of a Kite pattern. This chart suggests a person for whom evolutionary progress is a must. Baker devoted his life to exploring esoteric astrology and has published numerous volumes on the subject.

As in Alice Bailey's chart, Charon is found outside the close bundle of the other planets in Baker's Pluto-centered chart, as is typical in the majority of charts. Charon is sextile Jupiter and Venus, again reinforcing the Jupiter-Pluto connection and adding Venus's facility with con-

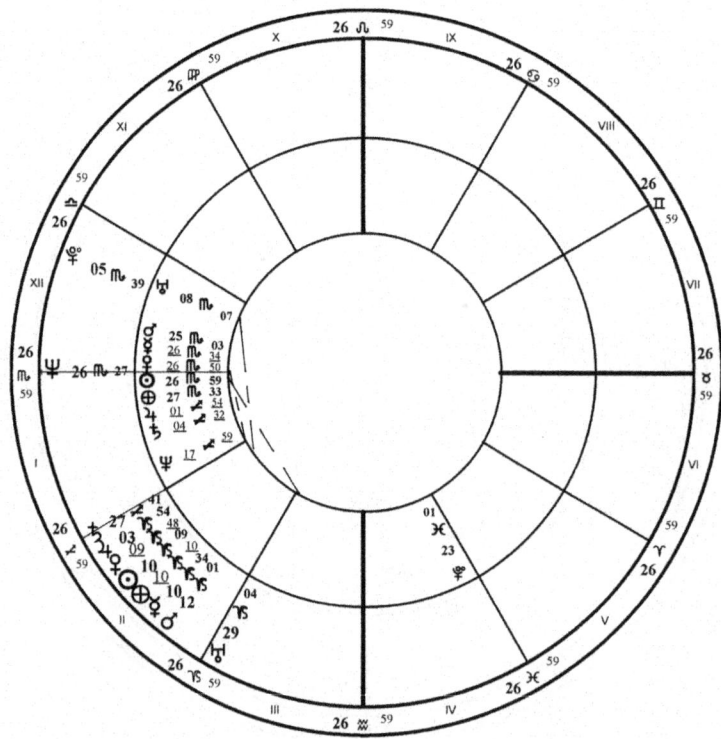

Comparison of Bailey and Baker Pluto-Centered Charts

crete knowledge. Dr. Baker took the body of esoteric work of Alice Bailey and wrote many volumes to clarify and elucidate points in organized, logical fashion.

How do the charts of Alice Bailey and Douglas Baker relate? In the Pluto-centered biwheel, we find Charon in Baker's chart not far from a conjunction to Bailey's Uranus. In addition, his Neptune is conjunct her Sun-Venus-Mercury-Mars-Earth. With only a seven-minute orb for Neptune to Mercury, we can conclude that Alice Bailey's voice, expressed through her writings, had a profound impact on Baker's life and writing.

In addition to the above aspects, Baker's Uranus is sextile Bailey's Earth, his Saturn semi-sextile her Mercury-Venus-Sun-Earth, and his Venus sextile her Uranus. This speaks to the intense psychic relationship between these two individuals. Baker's Jupiter is semi-sextile Bailey's Saturn, suggesting a connection of the level of wisdom and intelligent action as well. Baker's Saturn is semi-sextile Bailey's Earth, demonstrating that the two of them had a profound alliance of thought that allowed him to extend her work and make it available for people worldwide.

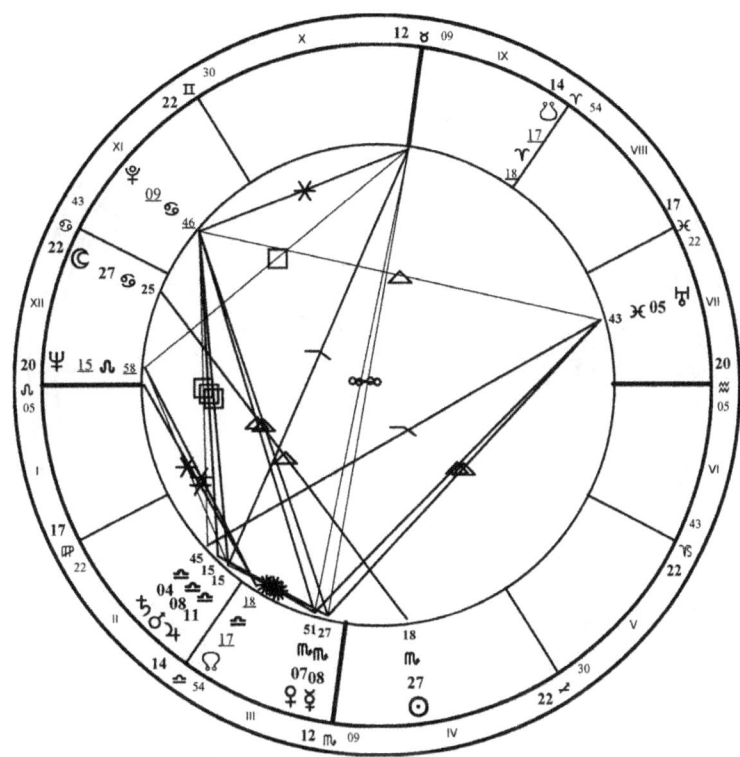

Roy Campanella Geocentric Birth Chart
Birth data: November 19, 1921, 10:45 pm EST, Philadelphia, Pennsylvania

Roy Campanella

I mentioned Roy Campanella in the chapter on Uranus because he was part of the magical Brooklyn Dodgers team that won the World Series in 1955. He was named most valuable player in 1955 and other years as well. His birth chart sports a Grand Trine of Pluto, Uranus, and Mercury-Venus. The closest square in the chart is from Jupiter to Pluto, so this is the energy that drives the chart. At the time of the opening game of the 1955 season, Campanella had solar arc Mars exactly opposition his Midheaven. In mid-September, transiting Pluto was trine his Sun. By the time of the World Series, transiting Mars was conjunct his solar arc Ascendant, and solar arc Saturn was creeping up to trine his birth Pluto.

The Pluto-centered chart includes a bundle of planets, with Neptune and Charon bracketing the tight inner planet grouping expected in Pluto-centered charts. Charon multiplies the energy of the chart by forming a very tight sextile to Saturn and a close square to Neptune, as well as a semi-sextile to Uranus. The Pluto-centered bundle shows how focused and intense Campanella

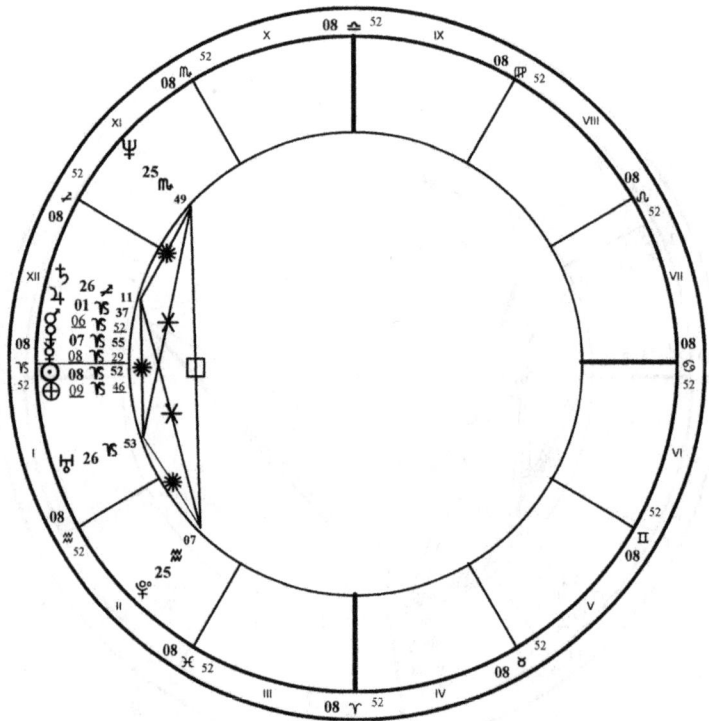

Roy Campanella Pluto-Centered Chart

was, and also suggests that focus is the reason for his great success as a ball player and businessman.

Reckless fool or magical adept? Pluto can be many things. Pluto is a powerhouse. The Pluto-centered chart points to the direction Plutonian energy takes, from the Pluto-centered perspective, thereby complementing the picture of Pluto in the geocentric chart. Sometimes the Pluto-centered chart seems to reveal very little, and other times it pins down the crazy magic of this distant, erratic planet.

Summary

1. Charon is more of a partner than a moon in size. How can this moon work for you as you wrestle with power and will?

2. Now the only planet outside the bundle around the Sun is Neptune, and even Neptune may be inside the bundle at times. The all-or-nothing feeling of power can be devastating.

3. Where is Charon by element and sign? How does Charon partner with the rest of the planets in the Pluto-centered chart to aid your quest for skillful means for using your power?

4. Does Charon aspect any of the planets? How do those aspects reveal the magic you have when will and intention guide your actions?

5. Charon transits the entire Pluto-centered zodiac in less than a week. For such a slow-moving planet, what does this suggest about how speedy Charon partners with powerful Pluto?

6. What does Charon tell you about your own approach to transformation?

7. What does Charon tell you about your attitudes toward death?

Whether astronomers consider Pluto to be a planet or not, the Pluto-centered chart reveals a rich lore for interaction as equals. Examination of your Pluto-centered chart may take only a few moments the first time around. Be prepared to get more information each time you return to this focused expression of power and will in your life.

Chapter Ten

Conclusion: Transpersonal Alchemy

The created universe, from its beginning, embodied basic goodness as an essential of being. This quality has not dwindled, nor will it ebb in the future.

Astrologers have the remarkable good fortune to be able to look into the heart of this basic goodness via the astrological chart. This particular volume has focused most directly on the planets as they relate to each other and on planet-centered astrology.

The very concept of a planet-centered chart has caused questions among astrologers ranging from, "I won't ever be on Mars, so why have a Mars-centered chart?" to "How can I add one more thing to what I need to learn?" to "Help me to learn about this new kind of chart NOW." These questions are certainly all valid. This work has endeavored to answer the first one, or at least provide food for thought concerning this new astrology. The second question is interesting in that we don't need to learn all that much new astrology in order to use planet-centered charts effectively. The third question comes from those curious enough to jump into a new area of study that promises to provide answers on all levels—physical, mental, emotional and spiritual.

Alchemical Associations of the Moons

Moon	Element	Effect/Style
Earth's Moon		Subconscious
Flores		Spiritual companion
Juno		Spiritual companion
Phobos		Action
Deimos		Thought
Io	Fire	Intuition, inspiration
Europa	Air	Intellect; insight
Ganymede	Water	Feeling; judgment
Callisto	Earth	Sensation; material perception
Tethys	Ether	Spirit
Dione	Fire	Intuition
Rhea	Air	Thinking
Titan	Water	Feeling
Iapetus	Earth	Sensation
Miranda	Ether	Spirit
Ariel	Fire	Intuition
Umbriel	Air	Thinking
Titania	Water	Feeling
Oberon	Earth	Sensation
Triton		
Nereid		
Charon		Spiritual companion

Mars and Neptune each have two satellites. Jupiter, Saturn, and Uranus each have numerous satellites, but each has four or five moons that are much larger than the rest. The roles of these three planets can be delineated through examination of the five satellites in terms of alchemical properties based on the four traditional elements and a fifth element, ether. Each of the three planets governs a major segment of our experience. Jupiter deals with process and with expansion and process; Saturn deals with concentration and structure; Uranus deals with change, intuition and the rhythm of life. The use of moons for each planet will reveal the fundamental alchemical reality of the three primary experiences of process, form and change.

By examining the planet-centered charts, an individual can obtain a much clearer view of personal processes and attitudes. Clarity is obtained by considering each chart as coming from the perspective of that planet, and not from one's personal perspective. If you have Saturn in Cancer in the second house, you have a personal set of opinions about Saturn. However, the Saturn-centered chart can delineate the transpersonal view of Saturn's role in your life. Instead of being limited by your own second house ideas, you can engage in the creative process of examining the possibilities from a more objective position—from Saturn's position. The five moons offer five specific alchemical possibilities for your consideration. You thus have left limitation behind and moved into a freer position of creative choice.

Planet-centered astrology offers a broad new vista of the philosophical or spiritual meaning of life.

Esoteric astrology predicted long ago that we would investigate large numbers of objects in our own solar system and this new approach considers many of them—the satellites of the planets. While a very strong beginning has been made here, there remains a great deal of research and contemplation. I hope to welcome many readers into the group of astrologers who pioneer this new ground, improving our understanding of astrology and improving our lives.

Appendix One:
Symbols for Moons and Asteroids

Planet	Asteroid or Moon	Symbol
Mercury	Flores	
Venus	Juno	
Mars	Phobos	
	Deimos	
Jupiter	Io	
	Europa	
	Ganymede	
	Callisto	
Saturn	Tethys	
	Dione	
	Rhea	
	Titan	
	Iapetus	
Uranus	Miranda	
	Ariel	
	Umbriel	
	Titania	
	Oberon	
Neptune	Triton	
	Nereid	
Pluto	Charon	

Appendix Two
Interpretations for Moons of the Planets and Two Asteroids

Planet	Moon	Planet Pairs	Intrepid Software Interpretation	Planet-Centered Interpretation
Mars	Phobos	MA/PL	Need for transformational action	Transformative Action
	Deimos	MA/MA	Need for immediate action	Inspired Action
Jupiter	Io	JU/SA	Far reaching objectives	Inspiration
	Europa	JU/UR	Optimism about accomplishment	Accomplishing action
	Ganymede	JU/JU	Excessive separation from family, urge to expand	Mirror-like wisdom
	Callisto	JU/NE	Mysticism, illusionary expansion	Equanimity
Saturn	Tethys	SA/ME	Logical thought	All-encompassing wisdom
	Dione	SA/EA	Resolution of chaos	Individual creative inspiration
	Rhea	SA/UR	Emotional tension or conflict	Learning from experience
	Titan	SA/SA	Multiple family integrations, concentration	Flow of creative process
	Iapetus	SA/MA	Endurance	Nature of satisfactory outcome

Planet	Moon	Planet Pairs	Intrepid Software Interpretation	Planet-Centered Interpretation
Uranus	Miranda	UR/UR	Multiple integration into community, consistency	Intelligent rhythm; ritual
	Ariel	UR/VE	Igniting of romantic love	Inspired change
	Umbriel	UR/ME	Insight	Focused thought
	Titania	UR/MA	Sudden, dramatic	Extreme feelings
	Oberon	UR/NE	Spiritual/ psychic states	Illusion of separation
Neptune	Triton	NE/NE	Exceptional separation from highly organized structure	Prophecy
	Nereid	NE/EA	Soul union	Compassion
Pluto	Charon	PL/PL	Exceptional separation from humanity	Individualized understanding
Venus	Juno	VE/VE	Harmony with many individuals or exceptional harmony with one person	Knowledge gained through interaction with others and the world
Mercury	Flora	ME/ME	Repeated application of intellect	Mediation/harmony via communication

Appendix Three
Houses in Planet-Centered Charts

Although I have not researched house placements in planet-centered charts extensively enough to arrive at a definitive statement about them, I have come to the conclusion that they have some meaning. The following are lists of considerations for each house. So far the houses appear to be the same for all planet-centered charts, and are similar to houses in solar charts, but *considered from the center planet's perspective.*

House 1: Sacral center, persona, fire that motivates you, anchor for potential, how you express this planet's energy most freely

House 2: Throat chakra, how this planet supports self-esteem, how you express this planet's energy

House 3: Brow chakra, spiritual communication, qualities of the spiritual teacher, intellect

House 4: Window into core beliefs, filter through which you view core beliefs, personal resources for spiritual growth, wellspring of ideas, inner voice of the daimon or personal angel

House 5: Chakra in the front of the throat related to the para-thyroids, mediator between higher and lower creative organs; creative activity of the consciously functioning soul

House 6: Heart chakra, service in the world, dharma

House 7: solar plexus chakra, partnership, enemies or challenges, how you relate to the world

House 8: Transformation, inner child, self-acceptance, qualities of transformation (sign here), influence of spiritual transformation (planets here)

House 9: Crown chakra, transcendent values or themes, faith, foresight

House 10: Spleen center, etheric and physical bodies and their interaction, effects of karma, expression or direction of dharma, purpose of this incarnation

House 11: Intuition, nature of rituals significant in your life, how you can best achieve equilibrium

House 12: Energies merging into consciousness, doubts, psychic insight, secret or hidden source of energy, compassion

Appendix Four
Eight Steps in Ritual Processes

We benefit from the steps in ritual process because we honor ourselves, others, and our actions. We sometimes get in trouble when we skip steps because we insult others or we create a lack of clarity.

The orderly steps of ritual process are as follows:

1. Purification: This could be as complex as fasting, prayer, and cleansing, or it could be as simple as taking a deep breath before beginning.

2. Centering: Meditation is often employed to achieve centering. However, an astrologer may simply view a relevant chart and become naturally centered in it.

3. Orientation: Alignment with geographic North provides orientation. Lining up properly for an activity does the same. An astrologer may include verification of accurate birth information and calculation as part of orientation.

4. Construction (any action word could be used here): Whatever the activity, this step reflects the content piece. For the astrologer, the consultation is the action of the ritual process.

5. Absorption: Now the people involved focus on what has just happened. They allow information to sink into their minds and they evaluate it.

6. Destruction: Well, we generally don't have to actively destroy anything. Saying goodbye at the end of a consultation breaks contact, thereby cutting the connection.

7. Reintegration: If the process has gone well, an earlier question or problem has been resolved. Now we take the acquired experience forward as part of our memory and understanding.

8. Actualization: The parties involved in the ritual will act out what they have learned in the future. Most successful rituals result in a sense of greater wholeness, optimism, or competence. As astrologers, we often can only evaluate our own actualization because the client has gone forward and may be out of contact.

In ongoing relationships, we move in and out of these steps naturally. Whenever we experience difficulty in relationships, we benefit from considering what may have been left out of a communication. Even when it is impossible to correct past errors, we learn more about how to do better in the future.

Appendix Five
Birth Data Sources

Chapter	Charts	Sources
One	Chart November 12, 2012	Future Date
Two	Abraham Lincoln	Rodden from *American Astrology*
	Martin Luther King	Rodden: Ruth Dewey quotes Dell 9/1970
	Barak Obama	Astrodienst
	Charles Darwin	Rodden, from Dell 4/49
	Clarence Darrow	Rodden from Church of Light, quoting *Astrological Review*
	William Jennings Bryan	Rodden, Ruth Dewey quotes *Star of the Magi;* Scopes Trial: *Wikipedia*
	Roman Catholic Church Edict	Church Edict: from Wikipedia 08/22/2009, "Evolution and the Roman Catholic Church"
Three	Elizabeth Taylor	Rectification by Carol Tebbs
	Johnny Cash	Rodden: Coleen Gauthier quotes his mother
	Adam Lambert	Lambert chart from various Internet sources (speculative time)
	Mata Hari	Astrodienst
	Greta Garbo	Astrodienst
Four	Broncos	From Bronco publicity Department
	John Elway	Elway birth time speculative
Five	Elisabeth Kubler-Ross	Rodden from Kubler-Ross to Robert Chandler
	Book publication	Publication Date, various Internet sources
	Sandra Day O'Connor	Astrodienst: birth certificate from Charley Settles
	Samuel Alito	Astrodienst: birth certificate from Charley Settles

Chapter	Charts	Sources
Six	R. Buckminster Fuller	Rodden: Frances McEvoy quotes him
	Tonya Harding	Astrodienst "C" rating
Seven	Arthur Young	From his book, *Nested Time*
	Brooklyn Dodgers	Time of start of first game
Eight	Timothy Leary	Rodden: birth certificate via Ed Steinbrecher
	Jonas Salk	Astrodienst from Beth Koch
Nine	Alice Bailey	Astrodienst: Rodden from Rudhyar
	Douglas Baker biwheel	Astrodienst: David Fisher quotes from Baker's book *Esoteric Astrology Part I*
	Roy Campanella	Astrodienst: Rodden, birth certificate

Glossary

Aphelion: the orbital position of a planet (moon) when farthest from the Sun (the planet it orbits).

Calcinatio: in western alchemy, any fiery process. The heating of the lapis in the purification process.

Coagulatio: in western alchemy, any earthy process. The process of solidifying the lapis in order to give it solid form.

Feeling: a function of consciousness whereby one can judge the value of an object.

Geocentric: Earth-centered.

Heliocentric: Sun-centered.

Intuition: a function of consciousness whereby one apprehends future possibilities.

Latitude: angular distance north or south from the equator of a body, measured in degrees.

Occultation: an eclipse of a star, planet, or spacecraft by a satellite or planet.

Perihelion: the point in orbit when a planet [moon] is nearest to the Sun [the planet it orbits].

Planet-centered Chart: a chart computed from the perspective of the planet itself, instead of the geocentric viewpoint. In each planet-centered chart, the Earth will be in the degree and sign OPPOSITE the planet itself in the geocentric chart, and all the planets will be placed according to the particular perspective of the center planet. Thus Venus will usually occupy different degrees in different planet-centered charts erected for the same time.

Sensation: a function of consciousness whereby one perceives the world directly through the senses.

Shadow: unknown and unacceptable dark components of the personality which remain unconscious until accepted by the individual.

Sidereal Period: the period for one object to complete an orbit around another body.

Synodic Period: the interval between successive similar lineups of a body with the Sun.

Thinking: a function of consciousness whereby one is able to recognize meaning or purpose in an observed object.

Vulcan: Name for planet thought to orbit between Mercury and the Sun. Esoteric astrologers assign Vulcan First Ray energy.

Zenith: the point directly overhead.

Bibliography

Almaas, A. H. *The Void: a Psychodynamic Investigation of the Relationship Between Mind and Space*. Berkeley, Diamond Books, 1986.

Alphabet. New York, Collier-Macmillan, 1966.

Arguelles, Jose, and Miriam Arguelles. *Mandala*. Berkeley and London, Shambhala, 1972.

Arroyo, Stephen. *Astrology, Karma and Transformation*. Reno, Nevada, CRCS, 1978.

Bailey, Alice. *Esoteric Astrology*. Lucis Pub. Co., 1951.

Bailey, Alice. *Treatise on Cosmic Fire*. Lucis Pub. Co., 1962.

Bennet, E. A. *What Jung Really Said*. New York, Schocken Books, 1966.

Besant, Annie, and C. W. Leadbetter. *Thought-forms*. Madras, India, The Theosophical Publishing House, 1969.

Bills, Rex. *The Rulership Book: A Directory of Astrological Correspondence*. Richmond, Va., Macoy, 1971.

Briggs, John and F. David Peat. *Looking Glass Universe; the Emerging Science of Wholeness*. New York, Simon & Schuster, 1984.

Bronowski, J. *The Ascent of Man*. Boston, Little, Brown Co., 1973.

Case, Paul Foster. *The Book of Tokens: Tarot Meditations*. Los Angeles, Builders of the Adytum, 1968.

Chogyam, Ngakpa. *Rainbow of Liberated Energy*. Longmead, England, Element Books, 1986.

Churchman, C. West. *The Systems Approach*. New York, Dell, 1979.

Circle Book of Charts. Comp. Stephen Erlewine. Ann Arbor, Mich., Circle Books, 1972.

Cirlot, J. E. *A Dictionary of Symbols*. New York, Philosophical Library, 1971. 2nd ed.

Claxton, Guy, ed. *Beyond Therapy*. London, Wisdom Publications, 1986.

Clement, Stephanie Jean. *Astrology of Development*. Tempe, American Federation of Astrologers, 2007.

Coudert, Allison. *Alchemy: The Philosopher's Stone*. Boulder, Shambhala, 1980.

Crowley, Aleister. *The Book of Thoth*. York Beach, Samuel Weiser, 1974.

Ebertin, Reinhold. *The Combination of Stellar Influences*. Aalen, Germany, Ebertin Verlag, 1972.

Edinger, Edward F. *Anatomy of the Psyche*. La Salle, 111., Open Court, 1985. Originally published in Quadrant, 1978 to 1982.

Fuller, R. Buckminster. *Synergetics*. New York, Collier-McMillan, 1975.

Graves, Robert. *New Larousse Encyclopedia of Mythology*. New York, Prometheus Press, 1968.

Green, Jeff. *Pluto, the Evolutionary Journey of the Soul, Vol. I*. St. Paul, Llewellyn Publications, 2000.

Hall, Manly. *The Secret Teachings of All Ages*. Los Angeles, The Philosophical Research Society, Inc., 1972 (1962).

Hamaker-Zondag, Karen. *Astro-Psychology*. Wellingborough, England, The Aquarian Press, 1980.

Harding, Esther. *Psychic Energy: Its Source and Its Transformation*. Princeton, N.J., Princeton University Press, 1963.

Hickey, Isabel. *Astrology: a Cosmic Science*. Watertown, Mass., 1970.

Hillman, James. *The Myth of Analysis: Three Essays in Archetypal Psychology*. New York, Harper Colophon Books, 1972.

Holmes, Ernest. *The Science of Mind*. New York, Dodd, Mead & Co., 1938.

Immegart, Glenn L., and Francis J. Pilecki. *An Introduction to Systems for the Educational Administrator*. Reading, Mass., Addison-Wesley Pub. Co., 1973.

Jacobson, Ivy M. Simplified Horary Astrology. Pasadena, Calif., 1970.

Jansky, Robert Carl. *Introduction to Holistic Medical Astrology*. Tempe, Ariz., American Federation of Astrologers, 1983.

Jansky, Robert Carl. *Planetary Patterns*. Venice, Calif., Astro-Analytics, 1975.

Jones, Marc Edmund. *The Guide to Horoscope Interpretation*. Wheaton, 111., Theosophical Publishing House, 1974.

Jung, Carl G. *Alchemical Studies*. Princeton, Princeton Univ. Press, 1983.

Jung, Carl G. *Man and His Symbols*. Garden City, New York, Doubleday & Co., 1964.

Jung, Carl G. *Mandala Symbolism*. Princeton, Princeton Univ. Press, 1972.

Jung, Carl G. *Mysterium Coniunctionis*. Princeton, Princeton Univ. Press, 1970.

Jung, Carl G. *Psychology and Alchemy*. Princeton, Princeton Univ. Press, 1968 (rev. ed.).

Jung, Carl G. *Psychology of the Transference*. Princeton, Princeton Univ. Press, 1969.

Jung, Carl G. *Synchronicity: an Acausal Connecting Principle*. Princeton, Princeton Univ. Press, 1969.

Jung, Carl G. *Visions Seminars*. Zurich, Spring Publications, 1976.

Kaldera, Raven. *MythAstrology: Exploring Planets & Pantheons*. St. Paul, Minn., Llewellyn Publications, 2004.

Kozar, Kenneth. *Humanized Information Systems Analysis and Design*. (Manuscript, 1988).

Laing, R. D. *The Divided Self*. Harmondsworth, England, Penguin Books, 1978.

Moacanin, Radmila. *Jung's Psychology and Tibetan Buddhism*. London, Wisdom Publications, 1986.

Ortega y Gasset, Jose. *Mission of the University*. New York, WW. Norton, 1944. (original address, 1930)

Parker, Derek, and Julia Parker. *The Compleat Astrologer*. New York, McGraw-Hili, 1971.

Ptolemy, *Tetrabiblos*. Hollywood, Symbols and Signs, 1976.

Rudhyar, Dane. *Astrology of Transformation*. Wheaton, 111., The Theosophical Pub. House, 1980.

Rudhyar, Dane. *New Mansions for New Men*. La Verne, Calif., EI Camino Press, 1978.

Rudhyar, Dane. *Practice of Astrology*. Boulder, Shambhala, 1978.

Ruperti, Alexander. *Cycles of Becoming*. Davis, Calif., Russell, Walter. The Secret of Light. Waynesboro, Va., University of Science and Philosophy, 1974.

Shelley, Percy Bysshe. *Works*. New York, Black's Readers' Service, 1951.

Small, Jacquelyn. *Transformers: The Therapists of the Future*. Marina del Rey, Calif., De Vorss & Co., 1982.

Smith, Edward W. L. *The Body in Psychotherapy*. Jefferson, N.C., McFarland & Co., 1985.

Synthesis 3-4: the Realization of the Self. Redwood City, Calif., Synthesis Press, 1977.

Tai Situ Rinpoche, Talk in Boulder Colorado, 1982, concerning the Ten Aspects of Knowledge (unpublished).

Trungpa, Chogyam, Rinpoche. *Mandala*. c. 1978 (transcribed lectures).

Trungpa, Chogyam, Rinpoche. *Myth of Freedom*. Berkeley, Shambhala, 1976.

Tucci, Giuseppe. *The Theory and Practice of the Mandala*. New York, Samuel Weiser, 1973.

Walsh, Roger, and Deane H. Shapiro, eds. *Beyond Health and Normality: Exploration of Exceptional Psychological Well-being*. New York, Van Nostrand, 1983.

Wilber, Ken. *Eye to Eye*. New York, Doubleday, 1983.

Wilber, Ken. *No Boundary*, Boston Shambhala, 1981a.

Wilber, Ken. *Up from Eden*. New York, Doubleday, 1981b.

Wolkstein, Diane. *Inanna: Queen of Heaven and Earth*. New York, Harper & Row, 1983.

Young, Arthur. *The Geometry of Meaning*. Mill Valley, Calif., Robert Briggs Assoc., 1976.

Young, Arthur. Nested Time: *An Astrological Autobiography*. Cambria, CA: Anodos, 2004.

Young, Arthur. *The Reflexive Universe*. Mill Valley, Calif., Robert Briggs Assoc., 1976.

Young, Ruth. *Guide to Arthur Young's Book, The Reflexive Universe*. (unpublished paper) Iain, C.C.. The Sacred Tarot. Los Angeles, Church of Light, 1979 (bound editions of lessons).

Endnotes

[1] Bailey, Alice. *A Treatise on Cosmic Fire*. New York, Lucis Publishing Company, 1973, pp. 793-795.

[2] The interpretations in this and other chapters are drawn from interpretations reports available in the Intrepid astrology program. These interpretation reports provide a systematic examination of the energy of each planet from it's own perspective and provide interpretations of each planet based on that viewpoint. You can view sample interpretations on the author's Web site at www.stephanieclement.8k.com or at Intrepid's Web site at www.aboi.com under SAMPLE PRINTOUTS.

[3] Zain, C. C. *The Sacred Tarot*. Los Angeles: The Church of Light, 1936, p. 80

[4] Rudhyar, Dane. *The Astrology of Transformation*. Wheaton, Ill.: The Theosophical Publishing House, 1980, pp. 68-69.

[5] Zain, *op.cit.*, p. 82.

[6] Bailey, Alice. *Esoteric Astrology*. New York, Lucis Trust, 1951, p. 67.

[7] Case, Paul Foster. *The Book of Tokens*. Los Angeles: Builders of the Adytum, 1972, p. 21.

[8] http://www.vatican.va/holy_father/benedict_xvi/speeches/2007/july/documents/hf_ben-xvi_spe_20070724_clero-cadore_en.html.

[9] Graves, Robert. *New Larousse Encyclopeida of Mythology*. New York, Prometheus Press, 1968, p. 130.

[10] Ibid., p. 17.

[11] Ibid., p. 19.

[12] Massey, Anne. *Venus: Her Cycles, Symbols & Myths*. Woodbury, Minnesota: Llewellyn, 2006.

[13] Bailey, Alice. *A Treatise on Cosmic Fire*. New York, Lucis Publishing Co., p. 298.

[14] Pottenger, Maritha and Zipporah Pottenger-Dobyns, *Healing Mother-Daughter Relationships with Astrology*. St. Paul, Minn: Llewellyn, 2003.

[15] Ptolemy, *Tetrabiblos*. Hollywood, Symbols and Signs, 1976.

[16] Graves, Robert. *New Larousse Encyclopedia of Mythology*. New York, Prometheus Press, 1968, p. 124.

[17] Hamaker-Zondag, Karen, *Astro-Psychology: Astrological Symbolism and the Human Psyche*. Wellingborough, Great Britain, The Aquarian Press, 1980, p. 157.

[18] Bailey, Alice, *Esoteric Astrology (Volume III, A Treatise on the Seven Rays)*. New York, Lucis, 1951.

[19] Ruperti, Alexander. *Cycles of Becoming*. Davis, Calif., 1978, p. 99.

[20] Jansky, Robert Carl. *Introduction to Holistic Medical Astrology*. Tempe, AZ: American Federation of Astrologers, 1983, p. 46.

[21] Bailey, Alice. *Esoteric Astrology* (Volume III of *A Treatise on the Seven Rays*. New York, Lucis Publishing Company, 1951.

[22] Fuller, R. Buckminster. *Synergetics*. New York: Collier-McMillan, 1975, pp. 272 and 296.

[23] Graves, p. 238.

[24] Kaldera, Raven, *MythAstrology*. St. Paul, MN, Llewellyn, 2004.

[25] Bailey, Alice, *Esoteric Astrology*; *Volume III of A Treatise on the Seven Rays*. New York, Lucis Publishing Co., 1979.

[26] Green, Jeffrey Wolf. *Pluto: The Evolutionary Journey of the Soul vol. I*. St. Paul, Minn., Llewellyn Publications, 2000.

[27] Bailey, p. 125.

www.ingramcontent.com/pod-product-compliance
Lightning Source LLC
Chambersburg PA
CBHW082235170426
43196CB00041B/2760